FORECAST

FORECAST

The Surprising—and Immediate—
Consequenses of Climate Change

STEPHAN FARIS

A HOLT PAPERBACK
HENRY HOLT AND COMPANY
NEW YORK

Holt Paperbacks
Henry Holt and Company, LLC
Publishers since 1866
175 Fifth Avenue
New York, New York 10010
www.henryholt.com

Library of Congress Cataloging-in-Publication Data
Faris, Stephan.
  Forecast : the surprising—and immediate—consequences of climate change /
Stephan Faris.
      p.   cm.
  Includes bibliographical references and index.
  ISBN: 978-0-8050-9084-0
  1. Climatic changes.   I. Title.
  QC981.8.C5F343 2008
  304.2'5—dc22   2008028558

Originally published in hardcover in 2009
by Henry Holt and Company

First Holt Paperbacks Edition 2009

Designed by Kelly S. Too

Printed in the United States of America
1   3   5   7   9   10   8   6   4   2

For Leonardo Geronimo

CONTENTS

FORECAST

INTRODUCTION

The first decade of this century will be remembered as the time when the world opened its eyes to climate change. Hurricane Katrina devoured New Orleans. The Amazon erupted in fire. Polar bears drowned in the melting Arctic. Heat waves swept Europe. Drought struck the American Midwest. Glaciers were melting like never before. In 2007, the Intergovernmental Panel on Climate Change, a coalition of scientists under the banner of the United Nations, declared that eleven out of the previous twelve years had been the hottest on record and that the warming had almost certainly been caused by emissions from our cars, factories, and power stations, and the clear-cutting of our forests. Later that year, the climate change panel shared the Nobel Peace Prize with Al Gore.

Life on Earth depends on a limited amount of greenhouse gases in the atmosphere. Naturally existing amounts of carbon dioxide, methane, nitrous oxide, water vapor, and ozone act like a blanket, trapping the heat from the sun's rays before it radiates back into space. Without them, the average temperature on the planet would be a bitter

2 degrees Fahrenheit below zero. The gases we have added since the Industrial Revolution are acting like an extra blanket. And we're busy piling on more. Since the beginning of the last century, the planet has warmed by roughly 1.2 degrees Fahrenheit. Before this one is out, the mercury will climb another 2 to 11 degrees.

But global warming will mean much more than hotter days. While some places are indeed seeing a rise in temperature, others are getting cooler. In some regions, global warming is clearing the skies of clouds. Elsewhere, it's overtopping the floodgates. Our rapidly changing climate has knocked the world off balance. Natural disasters— hurricanes, droughts, and forest fires—are striking harder and more frequently as it struggles to adjust. Nor will the impacts stop with the weather. Each climatic shift has effects that ripple down the causal chain. Droughts kindle competition over water. Heaving oceans put people on the move. Ecosystems shatter as their component species migrate or fail to adapt.

This book is about these types of impacts, the rolling series of events that reach beyond the environment and the weather to shape the way we live. In the time I spent on the research and travel for the book, I explored the impacts of our emissions on African refugee camps, Indian border towns, Mediterranean islands, European cities, Arctic outposts, Amazonian colonies, seaside towns on the Gulf Coast, and the vineyards of Napa Valley. What I discovered was that there are places in the world where impacts of the type we can expect from global warming have been experienced for a generation. In Darfur, a lengthy drought has pitted farmers against herders in a brutal conflict that has claimed hundreds of thousands of lives. Stronger hurricanes off the Atlantic and Gulf Coasts of the United States have made coastal cities more difficult places in which to live. Less alarmingly, increasingly warm summers have been a boon for the world's wine drinkers; grapes are coming off the vine riper and richer. And with the melting of the Arctic, shippers have begun to consider using polar routes to cut thousands of miles off their trips.

So far, much of the discussion about climate change has focused on the distant, catastrophic effects of a superheated world: cities deep underwater, frozen continents, the collapse of agricultural regions. But what I found was that even a modest amount of global warming—of the kind we can expect to experience in the next few decades—will be enough to set off dramatic shifts. Changes already under way point to impacts that range from the subtle and sometimes benign to the horrific and potentially catastrophic.

In the following pages I describe how disruptions in agricultural production can spark fighting, how the risk of fiercer natural disasters can upend coastal living, and how environmental refugees will raise the stakes in the immigration debate. I discuss how diseases will spread and become harder to contain, examine how the changing agricultural landscape will push farmers to adapt, explore how melting ice and failing rains threaten to redraw political boundaries, and look at the ways delicate regions risk slipping into catastrophe. We are already remaking—and will continue to remake—the globe's environment. Yet we don't have to guess at the consequences of a warming world. Its impacts lie just around the corner. The future of our planet can be found now, on the frontiers of climate change.

1

"THINGS WILL BREAK LOOSE FROM THE HANDS OF THE WISE MEN"
DARFUR, SCARCITY, AND CONFLICT

In 1985, with the Darfur region of Sudan deep in drought, a doctoral candidate named Alex de Waal met with a bedridden and nearly blind Arab sheikh named Hilal Abdalla. The elderly nomad and his tribesmen had pitched their camp across an unforgiving wasteland of rock and sand. Broad black tents rose like sails against the rough horizon. Thorn trees broke ground at lonesome intervals, sparse grazing for the tribe's camels. The student was long-limbed and gangly, bent forward with the eagerness of youth. The sheikh—tall, stately, stooped by age—asked him in. "His tent was hung with the paraphernalia of a lifetime's nomadism—water jars, saddles, spears, swords, leather bags, and an old rifle," De Waal recalled years later. "He invited me to sit opposite him on a fine Persian rug, summoned his retainer to serve sweet tea on a silver platter, and told me the world was coming to an end."

They dined on goat and rice and ate with their hands. De Waal was studying indigenous reactions to the dryness that gripped the region. The elderly nomad described things he had never seen before. Sand blew over fertile lands. The rare rain washed away alluvial soil. Farmers who had once hosted his tribe and his camels were now blocking their migration; the land could no longer support herder and planter both. Many of the sheikh's tribesmen had lost their stock and scratched at millet farming, relegated to sandy soil between plots of fertile land.

With his stick, the nomad sketched a grid in the sand, a chessboard de Waal understood to be the "moral geography" of the region. The farmers tended to their crops in the black squares, and the sheikh's people stuck to the white, cutting without conflict like chessboard bishops through the fields. The drought had changed all that. The God-given order was broken, the sheikh said, and he feared the future. "The way the world was set up since time immemorial was being disturbed," recalled de Waal, now a program director at the Social Science Research Council. "And it was bewildering, depressing, and the consequences were terrible."

Nearly twenty years later, when a new scourge swept across Darfur, de Waal would remember the meeting. Janjaweed fighters in military uniforms, mounted on camels and horses, laid waste to the region. In a campaign of ethnic cleansing targeting the region's blacks, the armed militiamen raped women, burned houses, and tortured and killed men of fighting age. Through whole swaths of Darfur, they left only smoke curling into the sky. At their head was a six-foot-four Arab with an athletic build and a commanding presence. In a conflict the United States would call genocide, he topped the State Department's list of suspected war criminals. De Waal recognized him. His name was Musa Hilal, and he was the sheikh's son.

• • •

On the path from worried elder to militant son lie the roots of a conflict that has forced 2 million mostly black Africans from their homes and killed between 200,000 and 450,000 people. The fighting in Darfur is usually described as racially motivated, pitting mounted Arabs against black rebels and civilians. But the distinction between "Arab" and "black African" in Darfur is defined more by lifestyle than by any physical difference: Arabs are generally herders, Africans typically farmers. The two groups are not racially distinct. Both are predominantly Muslim. The fault lines have their origins in another distinction, between settled farmers and nomadic herders fighting over failing lands. The aggression of the warlord Musa Hilal—forged in a time of desertification, drought, and famine—can be traced to the fears of his father and to how climate change shattered a way of life.

I first visited the region in the spring of 2004, when I traveled to the town of Adré on the eastern edge of Chad and drove three hours south along the Darfurian border. Refugees had been crossing the dry riverbed that formed the frontier and provided the area with its only source of water. It was early morning when I followed a group of women and their mules down into its dry sands. The rising sun made contrails from the dust at their feet. The vegetation was low and scrubby. The women dug broad holes into the bed of the river and pulled out brownish water with plastic jugs. It was dangerous work. The Janjaweed were still active on the other side and had been watering their horses and camels nearby. While their violence was mostly confined to Darfur, the militia had begun to launch cross-border raids in search of cattle that had escaped their assault.

The women had nearly filled their water barrels when two new arrivals dropped from the brush on the far side and fell to their knees in Muslim prayer. They were both women in their fifties, prematurely wrinkled and bent by sun and poverty. When they rose, I asked one to tell me her story. Halime Hassan Osman and her companion had left Chad the night before, risking discovery to return to Darfur and dig in the ruins of their village. They had spent the day in fear, hiding in the bush, too terrified to eat or pray, and returned after another night of fretful walking. The only thing they had found to salvage was a few handfuls of grains and beans. It had been a rough journey, and it seemed strange to me that these two grandmothers had been chosen for the mission. "If the men go, they will kill them," Halime answered. "If it's a young woman, they rape her. That's why it's us, the old women, who go see."

The refugees were camped in the high ground just inside Chad, in makeshift shelters of bent reeds and woven grass among the mud houses of a nearby village. On the wall of the local dispensary, the community's only concrete building, children had used charcoal to draw crude sketches of men and machine guns and planes dribbling bombs.

A fifty-five-year-old man named Bilal Abdulkarim Ibrahim showed me where he had been shot twice while saving his daughters from being raped. "I said, 'I will die. I cannot let you rape my girls in front of me.'" His attackers had tied a cord around his testicles and pulled, beating his wife when she tried to cut him loose. He only escaped when an older, white-bearded militiaman ordered his release. Fatum Issac Zakaria, a young girl, was seven months pregnant when she and three others were raped during an attack on her village. "We didn't want to go with them," she said. "They beat us all the way into the forest. They said, 'You are the wives of the rebels.' They insulted us. They said, 'You are slaves.'" That night as I was leaving, I saw flames rising from inside Sudan. They glowed for about ten minutes, then faded away. It was the Janjaweed, the refugees told me, torching the last of their homes.

Until the rains began to fail, the sheikh's people had lived amicably with the settled farmers. The nomads were welcome passers-through, grazing their camels on the rocky hillsides that separated the fertile plots. The farmers would share their wells, and the herders would feed their stock on the leavings from the harvest. But with the drought, the nomads ranged farther for their food, and the farmers began to fence off their land—even fallow land— for fear it would be ruined by passing herds. Sometimes they'd burn the grass upon which the animals fed. A few tribes drifted elsewhere or took up farming, but the camel-herding Arabs stuck to their fraying livelihoods—nomadic herding was central to their cultural identity.

The name *Darfur* means "Land of the Fur," called so for the largest single tribe of farmers in the region. But the vast region holds the homelands—the *dars*—of many tribes. In the late 1980s, landless and increasingly desperate Arabs banded together to wrest their own *dar* from the black farmers, publishing in 1987 a manifesto of racial superiority. It began with complaints of underrepresentation in the government and concluded with a threat to take matters into their own hands: "We fear that if this neglect of the participation of the Arab race continues, things will break loose from the hands of the wise men to those of the ignorant, leading to matters of grave consequences."

Clashes had broken out between the Fur and camel-herding Arabs, and in the two years before an uneasy peace was signed in 1989, three thousand people, mostly Fur, were killed, and hundreds of villages and nomadic camps were burnt. More fighting in the 1990s entrenched the divisions between Arabs and non-Arabs, pitting the pastoralists against the Fur, Zaghawa, and Massaleit, the three tribes that would later form the bulk of the rebellion against the central government. In these disputes, Khartoum often supported the Arabs politically. Sometimes—in an attempt to create a bulwark against revolutionaries from southern Sudan—the government provided arms.

When a rebellion began in Darfur in 2003, it was at first a reaction against Khartoum's neglect and political marginalization of the region. But while the rebels initially sought a pan-ethnic front against a distant, uncaring regime, the schism between those who opposed the government and those who supported it soon broke largely on ethnic lines. The camel-herding Arabs became Khartoum's staunchest stalwarts.

Nomadic Arab militia launched a brutal campaign to push the black farmers from Darfur. They wore military uniforms, sometimes drove military vehicles, and coordinated their attacks with Sudanese aerial bombing. Even so, the conflict was rooted more in land envy than in ethnic hatred. "Some of the Arab pastoral tribes, particularly the camel herders, did not have their own *dar,* so were always at the mercy of other tribes for land," said David Mozersky, the International Crisis Group's project director for the Horn of Africa. "This was fine for hundreds of years, since the system provided land for these groups as they moved. But as desertification worsened and as fertile land decreased, some of these pastoralists sought to have their own land, which wasn't really an option in Darfur. This was one of the main factors that Khartoum used to manipulate and mobilize these Arab tribes to join the Janjaweed and fight on their side. Interestingly, most of the Arab tribes who have their own land rights did not join the government's fight."

I returned to Chad later that year, to the refugee camp of Oure Cassoni, 170 miles north of Adré into the advancing desert. By then, the countryside on the other side of the border had been nearly depopulated. Most of those who were arriving had spent weeks fleeing the Janjaweed, seeking refuge in the hills, following the grass and water, trying to hold on to the last of their livestock, before they finally lost them and headed for the border.

Oure Cassoni was a place you'd go to only when you had nothing left. The surrounding desert was flat and featureless, cut by broad, shallow rivers that flooded when it rained and dried when it didn't. The trees spread out for water—low, thorny scrubs separated by wind-carved sand. Between them, nothing grew, not even dry grass. The horizon unfolded in a shallow arc, as if at sea. Dust devils spun themselves out against the noontime heat. The evenings blew in sandstorms from Darfur. They reared up as if to block the sun, turned everything sepia, then whipped down, leaving tents flattened and torn.

The nearest inhabitation, Bahai, was a rundown border town south of the camp, a scattering of concrete-block compounds with a population far smaller than its new, northern neighbor. "Bahai is a terrible place for a camp," Tim Burroughs, the environmental health officer for the International Rescue Committee, which was running the operation, told the *Christian Science Monitor.* "It's where the Sahara begins. There are plenty of dunes, you see houses overtaken by sand, you see villages abandoned." Groundwater was scarce. Aid workers drew water from a murky artificial lake at the border with Darfur, just three miles from the camp. Rebels were staged nearby and were rumored to be filling up their water tanks at the treatment center. In fifty years, Burroughs said, the area would likely be unable to sustain life.

The latest arrivals from Darfur had gathered outside the camp to wait for admission. The men wore robes of white or gray. The women were dressed in the incongruous colors of spring flowers. In the early morning, the refugees would tend to their mule if they had one and arrange their belongings if they had any. Everything slowed with the climbing sun. I moved from one group to another. The thorn-shaped leaves of the low trees offered little protection against the heat, so the refugees draped the branches with rugs, woven plastic mats, and empty rice bags. A small tree might shelter four or five people. No square inch of shade was wasted.

A sixty-five-year-old grandmother named Mariem Omar

Abdu had fled from her burning village three months earlier. She had watched from the bushes as men in military clothes shot three of her grandchildren. Their bodies were left where they fell. "We feared for ourselves," she said. "They killed our children. We feared the same would happen to us." She had watched as soldiers tied up three men from her village, beat them, loaded them on a truck, and drove them away.

Zahara Abdulkarim, a woman holding a small child, had large eyes with brows that curled up in a natural frown of concern. Her skin was smooth and unusually dark. Her robes were black. She had awoken one morning to the buzz of planes overhead, the bark of bombs from the far side of her village, and the rush of flames from inside her home. She ran out into her mud-walled courtyard and into the arms of the Janjaweed. One had a knife, and one had a whip. Both wore military uniforms. As they forced her to the ground, she saw her husband's body lying in the dirt. One held her while the other raped her. They called her a dog and a donkey, and when they were finished the man with the knife slashed deep across her thigh, a few inches above the knee. The mark signified slavery, they told her. She had been branded like a camel. "They want to replace this black skin with Arabs," she said.

Her analysis was echoed by the findings of Brent and Jan Pfundheller, retired law-enforcement officers who had spent a month in the camps conducting more than a thousand interviews in an attempt to quantify human-rights violations for the United States State Department. Men as well as women had been raped with sticks and rifle barrels and threatened with more of the same if they didn't leave. "One of the most common was, 'If you like this, stay in Sudan. If you don't like this, go to Chad,'" said Jan.

The attackers usually killed the men and boys, but allowed the women and girls to escape. Children were torn from their mothers' arms and checked for gender. "They'd take a girl, throw her to the ground," said Brent. "They'd take a boy, stab him. It takes a long time to make another generation of fighting age." Nobody was safe. The Janjaweed burned mosques and killed religious leaders. In one village, they burned five black imams alive. In another, invaders took a Koran from the mosque, threw it to the ground, and urinated on it. Men were killed at prayer, and one imam and his son were raped, then taken away. "We've worked in Bosnia and Kosovo, so there are certain things you expect," said Jan. "But I was shocked by the scope of the tragedy."

Why did Darfur's lands fail? For much of the 1980s and 1990s, environmental degradation in Sudan and other parts of the Sahel, the semiarid region just south of the Sahara, was blamed on the inhabitants. The dramatic decline in rainfall between the last forty years and the previous forty was attributed to mistreatment of the region's vegetation by the local population. Deforestation and overgrazing, the dominant theory went, exposed more rock and sand, which absorb less sunlight than plants, instead reflecting it back toward space. This cooled the air near the surface, drawing clouds downward, reducing the chance of rain, and continuing the cycle. "Africans were said to be doing it to themselves," said Isaac Held, a senior scientist at the National Oceanic and Atmospheric Administration.

But by the time of the Darfur conflict, scientists had identified another cause. Climatologists fed historical sea-surface temperatures into a variety of computer models of atmospheric change. What they discovered was that rising temperatures in the tropical and southern oceans, combined with cooling in the North Atlantic, was enough to disrupt the African monsoons and produce the changes in the weather that had been recorded. The degradation of Darfur's lands was a consequence, not a cause, of the drop in rainfall. "This was not caused by people cutting trees or overgrazing," said Columbia University's Alessandra Giannini, who led one of the analyses. The roots of the drying of Darfur, she and her colleagues had found, lay in changes to the global climate.

To what extent those changes can be blamed on human activities remains an open question. "Hurricanes rely on a similar pattern of warming," said Giannini. Just as it's possible to trace global warming through rising ocean temperatures to strengthening storms without being able to conclusively link a particular hurricane to increases in carbon dioxide in the atmosphere, so it is with Darfur. Scientists agree that greenhouse gases have warmed the tropical and southern oceans, and evidence indicates that sulfate aerosols from industrial pollution kept the North Atlantic cooler. But just how much man-made causes—as opposed to natural drifts in oceanic temperatures—are responsible for the drought that struck Darfur is as debatable as the relationship between global warming and the destruction of

New Orleans. "Nobody can say that Hurricane Katrina was definitely caused by climate change," said Peter Schwartz, the coauthor of a 2003 Pentagon report on climate change and national security. "But we can say that climate change means more Katrinas. For any single storm, as with any single drought, it's difficult to say. But we can say we'll get more big storms and more severe droughts."

Darfur may be a canary in the coal mine, a foretaste of climatically driven political chaos. Even mild climatic shifts over the past thousand years, the kind that have occurred naturally, seem to have the power to spark conflict. David Zhang, a professor at the University of Hong Kong, scoured China's dynastic archives for records of war and rebellion and compared them with the historical temperatures in the northern hemisphere as gleaned from analysis of tree rings, coral, boreholes, and ice cores. In the span between the eleventh and twentieth centuries, Zhang and his colleagues counted fifteen periods of intense fighting. All but three of them occurred in the decades immediately following extended periods of unusual cold. The most dramatic peak in warfare occurred when China—and Europe—was plunged into the Little Ice Age. "At that time, globally, you see a very, very sad situation," said Zhang. "In Europe, you had the Seventeenth-Century Crisis, you had the Thirty Years' War."

In China, Zhang had found, cooling temperatures reduced agricultural yields, leading to famine, rebellion, and war. "We also found that dynastic collapses also followed the oscillating temperature cycles over the past millennium," Zhang wrote in the scientific journal *Human Ecology*. "Almost all of the dynastic changes occurred in cold phases, with the exception of the Yuan dynasty, which collapsed eight years after the end of a cold phase, although it had lost most of its territory in the 'Late Yuan' peasant uprisings during the cold phase. The delayed collapse was largely a result of power struggles among different rebel groups." But what rattled Zhang most was the scale of the historical climatic change as compared to the warming the world has experienced this century. "If you look at the average temperature, it was 0.3 degrees centigrade cooler [than the historical average]," he said. "Nonetheless, the impact is so big. Right now we're 0.7 centigrade higher than the average. I don't know what's going to happen."

The effects of global warming will be felt all over the world, often in unexpected ways and surprising places. "Sudan's tragedy is not just the tragedy of one country in Africa," said Achim Steiner, director of the United Nations Environment Programme. "It is a window to a wider world underlining how issues such as uncontrolled depletion of natural resources like soils and forests allied to impacts like climate change can destabilize communities, even entire nations. It illustrates and demonstrates what is increasingly becoming a global concern. It doesn't take a genius to work out that as the desert moves southwards there is a physical limit to what ecological systems can sustain, and so you get one group displacing another. Societies are not prepared for the scale and the speed with which they will have to decide what they will do with people."

It's difficult to know where global warming will strike hardest, so the best way to predict where climate-induced conflicts might break out is to identify the countries least able to withstand the stress. "It's best not to get too bogged down in the physics of climate," said Nils Gilman, an analyst at Global Business Network, a strategic consultancy based in San Francisco, and the author of a 2007 report on climate change and national security. "Rather you should look at the social, physical, and political geography of regions that are impacted. It's not like Darfur would have been a happy part of the world if not for climate change." Tensions between herders and farmers existed long before Darfur began to dry. With climate change they erupted into war.

Early hotspots are likely to be in Africa south of the Sahara and in places like Central Asia or the Caribbean, where institutions are weak, infrastructure is deficient, and the government is incompetent or malevolent. International Alert, a group that focuses on conflict resolution, has compiled a list of forty-four countries that climate change will put at high risk of armed conflict, including Iran, Indonesia, Israel, Algeria, Nigeria, Somalia, Bolivia, Colombia, Peru, and Bosnia and Herzegovina.

The crisis in Darfur has already spilled over into Chad and the Central African Republic. Nomads from Sudan are pushing deep into the Congolese rain forest. When the United Nations Security Council held its first-ever debate on the impacts of climate change in 2007, the Ghanaian representative stood up to declare that he hoped the "repeated alarm" about the threats posed by global warming would "lead to action that is timely, concerted, and sustainable." In his country, he said, nomadic Fulani cattle herdsmen were buying high-power assault rifles to defend their animals from angry local farmers. Climate change had expanded the Sahara desert, forcing the pastoralists into agricultural lands.

As global warming threatens to push countries all over the world into conflict, those looking to head off future crises will want to know what volatile mix of land pressures and local politics will push a tense region over the edge. "The first thing you do is look around the world for regions where you have a very large population of people who are directly dependent on limited resources: local cropland, water supplies, forest," said Thomas Homer-Dixon, a political scientist at the University of Waterloo who has studied the links between environment and conflict for nearly two decades. "Then you look for places where the resources are already severely degraded. Water is really scarce. Cropland is damaged. Deforestation is widespread."

"The next thing you look for is places where the governments and social coping mechanisms are in real trouble," he said. "These are places where governments are corrupt, where ethnic divisions are already deep, where capital is limited, where there is poverty. That combination—of corruption, deep ethnic division, and inadequate capital—creates enormous vulnerabilities. You add climate change on top of that? Bingo. You've got a complex crisis."

HAITI'S PROBLEMS ARE ONLY TANGENTIALLY LINKED TO GLOBAL warming, but a comparison with its neighbor offers a glimpse at the challenges Darfur and other countries like it might face in an environmentally stressed future. The Dominican Republic shares the troubled country's island, but not its endemic deforestation and erosion. While Haiti has lost 98 percent of its forest cover, the Dominican Republic has largely managed to preserve its trees. The difference is visible not only from the air, where the border is demarcated with an abrupt shift from lush green to bare brown, but also in the two countries' death tolls. When Hurricane Jeanne struck the Dominican Republic in 2004, it killed eighteen people. In Haiti, where the storm didn't even make landfall, more than three thousand lives were lost under floodwaters and mudslides. Deforestation had left the slopes too weak to be able to retain the downpour.

Global warming is likely to make Haiti's situation worse, as rising variability in the weather means more floods and more droughts. Stronger rains will wash away fields, roads, and buildings. Failing crops will bring ruin to the countryside and chaos to the cities. In April 2008, rising food prices, fueled in part by climate change, sparked unrest in the country's urban areas. The cost of basic staples—beans, rice, milk—had jumped by 50 percent, exhausting the patience of a country where the poorest regularly ease their hunger with meals made of mud. Protests turned to riots. Shops were looted. Cars were burnt. At least six people were killed, including a United Nations peacekeeper from Nigeria shot while trying to contain the violence. After a week of chaos, lawmakers voted to dismiss the country's prime minister, claiming he had lost the confidence of the electorate.

But, even without the added burden of climate change, Haiti would likely have remained much poorer than its neighbor. Its environmental problems have become entangled in its political woes, making both harder to fix. Decades of poverty, population growth, and near anarchy have stripped the countryside of its forests and split farms into small, infertile plots. "What you see in Port-au-Prince—the concentration of people in the slums, which creates violence, which creates disease—it's because the people cannot produce more in the countryside," said Max Antoine, executive director of Haiti's Presidential Commission on Border Development, tasked with reforesting the area near the Dominican Republic. "So they leave their lands and come to the city hoping to find a better life. And of course they can't find a better life. So what do they do? They have to eat. So they start being gangsters. They become susceptible to drug dealers."

To get a look at the challenges under which the country is struggling, I put the slums at my back and drove out of Port-au-Prince. The road had been recently worked on, and we quickly overtook pickup trucks packed with passengers huddled against the dust. At the city's farthest-flung gas station, five horses lay in the road, their bellies swollen with death. I held my breath as we eased around them. I had wound my window down against the warmth and did not want to find out if they smelled as bad as they looked.

Just outside of town, the tropical hills had the look of a desert landscape, smooth and nearly treeless, broken in white gashes where the land had given away. It was as if natural selection had favored the short or spindly, not so much as a way to survive the heat as much as to escape the machete. Where the road had been widened, fresh cuts in the hillside revealed one cause of the country's woes. In cross sections that would have towered over a freight truck, only the top few inches of the strata exposed brown, fertile soil. With nothing but dangling grass roots to anchor it to the dusty, white limestone, it wouldn't have taken much to wash the earth away.

What makes Haiti's problems so intractable is the complex and painful ways they feed on each other. The impoverished country depends on trees for 71 percent of its energy use: firewood in the countryside, wood charcoal in the cities. For an impoverished peasant, stripping the forests has become a way to get by. "If I'm a farmer and my crops are failing, what can I do?" said Antoine. "Do I die today? Or do I extend my life for the next few days by cutting trees and selling charcoal so I can buy medicine? So I can buy some fertilizer so I can grow some lettuce?" When the forests are gone, the slopes can't hold on to their soil. Entire villages are lost to mudslides. Roads and bridges are damaged. The slums continue to swell. Haiti sinks deeper into poverty. Pressed to survive, another farmer chops down another tree. "It's not a vicious circle," said Philippe Mathieu, the Haiti director for the Canadian charity Oxfam-Québec. "It is a spiral. Each time you make a turn, you have less space."

The road rose sharply, and we were at our destination, Lac de Péligre, an artificial lake stretching ten miles towards the Dominican border, held back in a narrow valley by a towering concrete dam. Completed in the 1950s, the hydro-power plant had been built without consulting or really even informing the farmers whose fertile fields and orchards would soon be at the bottom of a lake, and it had devastated the local community. "One of the old people of Cange remembered seeing the water rising and suddenly realizing that his house and goats would be underwater in a matter of hours," wrote Tracy Kidder in *Mountains Beyond Mountains.* "'So [the man said] I picked up a child and a goat and started up the hillside.'"

"Families had hurried away, carrying whatever they could save of their former lives, turning back now and then to watch the water drown their gardens and rise up the trunks of their mango trees," Kidder continued. "For most, there was nothing to do but settle in the steep surrounding hills, where farming meant erosion and widespread malnutrition, tending nearer every year toward famine."

The builders of the dam had left their construction equipment behind. Skeletal structures of cement and steel towered on either side. In a dusty market square, an abandoned construction crane reached for something above the trees. Villagers used its bulk to tie off their horses or spread their laundry across its raw metal treads. I walked out onto the concrete stretch that separated the lake from the river below. Ferrymen paddled carved pirogues, their passengers sunk dangerously below the waterline. Over the dam's edge, the steep drop looked down upon the blue, rectangular warehouse containing the power-generating turbines. Pylons marched in single file over a low hill toward the capital. Violent white foam twisted and bucked on its way to join the riverbed.

"The project was intended to improve irrigation and to

generate power," Kidder wrote. "It wasn't as though the peasants of the central plateau didn't need and want modern technology. . . . But, as they themselves often remarked, they didn't even get electricity or water for their land. Most didn't get money either. In fact, the dam was meant to benefit agribusinesses downstream, mostly American-owned back then, and also to supply electricity to Port-au-Prince, especially to the homes of the numerically tiny, wealthy Haitian elite and to foreign-owned assembly plants."

Had it at least accomplished that, the project might have generated economic growth. Instead, the water I was watching was being wasted. At full capacity, the plant produces sixty megawatts of power, enough to supply a small city. But deep in its bowels below, the needle hovered between fifteen and seventeen megawatts. Of the plant's three truck-sized turbines, two were broken, disassembled and waiting for pricey spares or improvised replacements. The entire operation relied on a single turbine—still in use despite a broken seal that spit water into the work space and a defective shaft that wobbled worryingly as it spun. With the reservoir near capacity, the dam was evacuating water, churning away the force that could be used to produce electricity.

The dam had been planned to last 140 years. But the engineers who projected its life span had assumed the valley's wooded hills would become more forested. They hadn't foreseen that the dislocated farmers would settle on the steep slopes, cutting the trees to survive, or migrate to the capital to become another mouth to feed with charcoal-cooked meals.

I had been joined on the dam by two of its security guards. Every slope I could see was flattened by clear-cutting. Patches of pale green grass hung above steep, scraggly fields that slipped precariously towards the water. The forest clung only where the slopes were steepest. "People need to eat," said one of the guards. He made a quick two-handed gesture to his stomach. For those in desperate need of cash, even fruit trees were worth more as charcoal than as orchards. "They cut the mango, and then they sell the charcoal," said the other. "As the money finishes, the people return into misery. Then they go and cut another mango. And it's like that that the deforestation happens."

Rising silt in the reservoir had shifted the currents near the dam and warped a protective shield meant to keep branches and logs from jamming the system. Sediment levels hadn't been measured since 1988. But at some point in the next few decades, unless hundreds of thousands of dollars are spent for dredging, the reservoir will simply fill up. "It's another vicious circle," Joanas Gué, Haiti's agriculture secretary, told me when I visited him in his offices on my return to Port-au-Prince. "Not only do we have a reduction of our potential energy production, we have to spend money to undo the damage of the deforestation."

"Do you know how many trees are cut in Haiti each year?" he asked. "Thirty million trees. That's a lot." The government had planned an ambitious program of planting 140 million saplings over the next five years. But given the speed of cutting, even that furious pace would be nothing more than treading water in the middle of the ocean. "Even if we do a big reforestation program with fruit trees, mango trees, avocado trees, they'll be cut one day or another," he said.

"You have to do everything at once," Gué said. "If you plant trees, but don't provide another source of energy and forbid people from cutting trees to make charcoal, demand will rise for an alternative that we don't have yet. The price of charcoal will increase: we'll have contraband. If, on the other hand, we provide an alternative source of energy to the people of Port-au-Prince and we don't give another source of revenue to the peasants, at that moment, the price of charcoal will drop. They'll have to cut a lot more trees to satisfy their needs in terms of monetary revenue." Haiti's cash-strapped government is unable even to keep order in the streets of the capital. Yet to solve its country's problems—that is to raise standards to the levels of the Dominican Republic, itself a poor country—it will have to solve three puzzles at the same time, any one of which seems beyond its capabilities. "We need to provide a revenue source," said Gué. "We need a planting program. And we need to create an alternative source of energy."

With weak governments vulnerable to weak shocks, preventing future conflicts like Darfur's will mean keeping countries from sliding into these sorts of traps. Where rains fail, drought- and salt-tolerant crops will improve day-to-day living. Better roads will ease access during emergencies. Reservoirs and irrigation can carry communities through droughts or absorb flooding. "Basically what you want to have is more water control," said Claudia Ringler, a water management specialist at the International Food Policy Research Institute in Washington, D.C.

In Darfur itself, where the crisis has spiraled too far down for a short-term fix, recognizing climate change as a player in the conflict means seeking a solution beyond a political treaty between the rebels and the government. The United Nations Environment Programme predicts the desert will continue to expand into Sudan's farmland, cutting agricultural production. Rainfall is expected to continue to decrease. "The Darfur conflict began as an ecological crisis, arising at least in part from climate change," wrote Ban Ki-moon, the United Nations secretary-general, in a 2007 op-ed in the *Washington Post.* "It is no accident that the violence in Darfur erupted during the drought."

"Any peace in Darfur must be built on solutions that go to the root causes of the conflict," he continued. "We can hope for the return of more than two million refugees. We can safeguard villages and help rebuild homes. But what to do about the essential dilemma—the fact that there's no longer enough good land to go around?"

Haiti's experience with deforestation shows how environmental pressure can undermine a society's ability to cope with its effects. In Darfur, fighting over failing lands makes it nearly impossible to rethink land ownership or management. With most of the population in refugee camps, rebel groups are fighting each other and attacking peacekeepers. The country has slipped into anarchy. Unless Khartoum suddenly decides to cooperate, creating the conditions for negotiations will require forceful intervention and a long-term stay. "The chance of finding new ways of reforming land management during a time of conflict is pretty much zero," said Homer-Dixon. "The first thing you've got to do is stop the carnage and allow moderates to come to the fore." It's a heavy commitment, one made more burdensome by Darfur's harsh environment. In 2007, United Nations officials were struggling with the logistics of deploying an estimated twenty-six thousand peacekeepers in an area where humanitarian groups could barely provide refugees with a few quarts of water a day. Each soldier required twenty-two gallons a day. The mission was planning twenty daily flights just for water.

They better find the funds to keep those planes flying for years. To craft a new status quo, one with the moral authority of the God-given order mourned by Musa Hilal's father, local leaders will have to put aside historic agreements and carve out new ones. Lifestyles and agricultural practices will likely need to change to accommodate many tribes on more fragile land. Widespread investment and education will be necessary. "Solutions imposed from the outside rarely graft," said Homer-Dixon. "Local solutions to locally generated conflict will last longer. But these processes can take decades."

The impact of climate change on a country is analogous to the effect of hunger on a person. If a starving man succumbed to tuberculosis or was shot while stealing a piece of bread, you wouldn't say he had died because he didn't eat. But hunger played a role in his death. Global warming by itself doesn't launch wars, rebellions, or campaigns of ethnic cleansing. "What climate change does is decrease the resilience of a society," said Homer-Dixon. "It makes it more brittle and more vulnerable to shock and various kinds of pathologies, including major violence."

Of all the repercussions of climate change on the killing in Darfur, one of the most significant may be moral. If the region's collapse was in part caused by the emissions from our factories, our power plants, and our cars, we bear some responsibility for the dying. "This changes us from the position of Good Samaritans—disinterested, uninvolved people who may feel a moral obligation—to a position where we, unconsciously and without malice, created the conditions that led to this crisis," said Michael Byers, a political scientist at the University of British Columbia. "We cannot stand by and look at it as a situation of discretionary involvement. We are already involved."

2

"WE'RE THE FAR COUNTRY"

THE GULF COAST, WARMING WATERS, AND THE FLIGHT FROM PARADISE

It was a sunny summer afternoon off the southern Florida coast. The wind was in our faces, and our small boat bucked over low waves. Jet Skiers left contrails across the chopping water. The boat's captain was a former actor named Richard Grusin who had played Tom Cruise's wrestling coach in *Born on the Fourth of July*. He wore a blue tank top over a black swimsuit. His feet were bare and callused. He stopped the boat in a patch of calm. The island of Key West was a rusty smear on the coin's edge of the horizon. The water was turquoise and blue, mottled in brown and yellow by the reef he had brought me to see.

Before setting out, Grusin had told me that all but 6 percent of the coral in the nearby marine sanctuary had been lost, that the summer water temperature regularly hovered between eighty-eight and eighty-nine degrees, that one day, earlier in the season, when he pulled into the reef, the thermometer read ninety-one degrees. "I've never seen it that high," he had said. "You're going to go in the water. You won't believe it. It's like a bath."

I flapped up to the side of the boat in my fins, pulled on

my mask, and asked what I should look for. "It'll be pretty evident," he said.

I dropped into a scattering of yellowtail snapper, took my bearings, and kicked out towards the coral. The nearest spur was concrete gray with tiny patches of yellow. Here and there, a purple fan coral waved in the current. The occasional crevice offered a patch of labial pink or orange, but most of the coral was dead, littering the floor like an underwater ossuary. Small living colonies, none of them bigger than my arm, spread out like lichen on a sidewalk. Schools of snapper dodged hefty rainbow parrotfish. It took several minutes of swimming in larger and larger circles before I understood that this was all I was going to see.

Scientists monitoring the reef say that the warming waters around the Keys have led to large-scale bleaching. Coral colonies, stressed by unusual temperatures, expel the symbiotic algae that give them their radiant colors, leaving them bone white and weakened. Sometimes they recover, reabsorbing the algae. But more often it's the first step towards death. When Grusin had spoken about the 6 percent of living reef, I had pictured a thriving island of coral, fish, eel, and anemones set in a wasteland of abandoned rock. Instead, the 6 percent was peppered across the seabed. In no place was it thriving.

I was joined in the water by Grusin's first mate, a onetime hard-hat diver named David Pasquale, or Old Gray Dave, who first worked Key West in the 1970s and 1980s and had only recently returned. We kicked down together, dove maybe fifteen feet to look for fish under an overhang, found nothing, and arched back up. When we first met, I had asked him how much the reef had changed in his absence, and he had given a realistic but upbeat answer. "I'd call it a day-and-night difference," he said. "But it's the best you're ever going to see in your lifetime. So try to enjoy that." Surrounded by so much gray devastation, I was finding it hard to take his advice. As we broke the surface, I pulled off my mask.

"Even just seeing a bit of color is rare," I said.

"I know," he said. "When I first came after twenty years, I almost cried. They'd ask me, 'In a word, how would you describe the reef?' I'd say, 'In a word, how would you describe Hiroshima?'"

I stuffed the snorkel into my mouth and started paddling towards the boat. The sea was warm enough to be not at all refreshing, and I was tiring quickly.

The coral colonies may be among the first victims of the warming waters, but they aren't going to be the only ones. If the reefs vanish, the fish will follow. Divers might be placated by purposely sunk wrecks, but snorkelers and sport fisherman will have little reason to visit. Nor will the impact of climate change be constrained to underwater ecosystems. Sea levels will rise, gently at first, then perhaps dramatically. Hurricanes will be stronger and more dangerous. "The Florida Keys have developed a pattern of tourism that has a rhythm to the season," said Jody Thomas, the Nature Conservancy's director for the Southern Florida Conservation Region. "That's going to get disrupted. What's that going to mean to the economics of the area? This area is sort of ground zero in the developed United States. Things are going to happen here first with climate change, and they're going to happen dramatically."

Key West has always drawn its fortune from the sea, booming and busting through waves of opportunistic industry and benign lawlessness. It's the ultimate frontier town, the southernmost city in the continental United States. Connected to mainland Florida since 1912, first by rail, then by road, it dangles like the last pearl on a band of marshy islands strung into the Caribbean. The island has yielded riches from ship salvaging, cigar rolling, and sponge harvesting. It stakes out the strategic seaways where the Gulf Coast joins the eastern seaboard. Shipping lanes stretch south past Cuba towards the Panama Canal. During Prohibition, Key West served as a center for bootlegging. World War II transformed it into a military town, the civilian satellite of an adjacent navy submarine base. In the 1970s, it turned to drug and people smuggling.

The town has built its latest boom on tourism. A longtime gay getaway, Key West has opened its doors to the mass market and the resort class both, capitalizing schizophrenically on its reputations for hidden-away debauchery and peaceful remoteness. It's a tropical island vacation you can drive the kids to. It's drinking, drugs, and nudity. It's sun and beach, family and fun, scuba diving and parasailing. It's a destination of no restraint, where the only thing out of bounds is a sense of shame.

The one-mile walk down Duval Street, the town's main drag, begins with the Gulf of Mexico at your back. There are Mexican bars, Irish bars, Cuban bars, sports bars, cocktail bars, just plain bars, and an upstairs strip bar. Glass window cases display hats and pipes, bongs and thongs, T-shirts stamped "Marilize Legjuana" and underwear boasting "I ❤ to Fart." About halfway across town, the mood begins to turn. Just beyond the Garden of Eden—the town's "clothing-optional" drinking hole—a Starbucks faces a Banana Republic. Coral and shell fills the store windows. An establishment featuring singing drag queens offers the street's last touch of debauchery and serves as the gateway to the town's gay district, a strip of stately B and Bs, tony bistros, and meticulously kept cafés that dominates the Atlantic side of the boulevard. Here and there, tucked away from the crowds, a Hyatt or a Westin lowers its gates and turns its back to the mass market and its face towards the warming waters of the Caribbean.

"It's just become alarming to everybody," said Grusin. "It used to be that the blue water would come right up to the shore. Now for a lot of the summer, it's green. And it's green because of the nutrients in the water and because of the pollution and the temperature. Algae feeds on the nutrients. So unless the Gulf Stream comes in, the blue water is hardly here anymore."

"A lot of people who work around the water are starting to get staph infections," he said. "I had one. Dave almost died from one."

"I got sun poisoning on my lips and they cracked," said Pasquale. "And I continued working. A lump started coming up on my neck. And one day Richard said, 'You look terrible; what's going on?' And I said, 'I feel terrible.'" Pasquale told his bosses he was taking a couple of days off, drove to his wife's work, and fell asleep in his truck waiting for her to come off her shift. "When she came out, the whole side of my neck had swelled up like a football," he said. "My lymph nodes were like a mountain range. They rushed me to the hospital, and I said, 'I just want to go home and go to bed.' And they said, 'You would never wake up.'"

"Big shots and big antibiotics," he said. "I was off work for almost six weeks. Getting a glass of water made me tired. I'd have to sleep for hours. I'd wake up and go to the bathroom, and that would make me tired. The woman who worked on me said they're getting five to seven cases a day."

Jody Thomas's Key West home lies just across the street from the town cemetery, the island's highest point, eighteen feet above sea level. The Nature Conservancy has joined with the National Marine Sanctuary and others to study coral bleaching in hopes of finding ways to mitigate or adapt to the stresses of climate change. But increasingly they're turning their attention to what lies onshore. Chris Bergh, the director of the Conservancy's Florida Keys program, rolled out a laminated poster across Thomas's wooden floor. On it was a large elevation map of Big Pine Key, a low-lying island up the Keys where the Nature Conservancy is a partner in a deer refuge and where Bergh has his home. Across the world, the temperature at the surface of the sea rose nearly one degree Fahrenheit during the twentieth century. As the waters continue to warm, they will expand. Four smaller maps showed the same island under different levels of rising water. The first depicted the expected sea level rise by the end of the century under the most optimistic conditions, a scenario that assumes the world cooperates to aggressively fight climate change. The seven-inch rise had flooded 16 percent of the island. Water sloshed through some of the streets and presumably the ground floor of many houses.

The next two maps were based on more realistic assumptions, and the sea rose accordingly. Under a scenario of rapid world economic growth with little attention to climate change, the track the planet is on now, waters had risen two feet. Waves washed over more than half the island. Bergh's final map reflected the work of the German climatologist Stefan Rahmstorf, who junked the computer models and simply projected current trends to the end of the century. Big Pine Key was almost entirely under the sea, just a grid of rooftops and a scattering of highlands rising above the waves. "The important thing is that we don't know where we're going to fall on this spectrum," said Bergh. "And storm surges and abrupt changes could make it happen much more quickly."

"We're having conversations that we never dreamed of having," said Thomas. "The Nature Conservancy has got to start saying, 'Okay, what's our investment worth to us, and should we continue investing?'"

"We're not buying land here anymore," said Bergh. "Where sea levels have been rising, there are winners and losers. It's certainly worse for us who live on the island. And it's worse for the terrestrial species. But all that then becomes marine environment. Once this is underwater, we still care about it. It will still be part of the Nature Conservancy's mission when there's coral there and fish there rather than deer and plants. If we can take the energy and money that we would have spent buying land to prepare the terrestrial environment to be a good future marine environment, that's a really worthwhile thing to do. So if we have a toxic waste dump, get it cleaned up now before it goes underwater and spills out into the future marine environment."

"Our strategy is to use this place as an example of why we should try to minimize the impacts of climate change and minimize the inputs that create it," said Thomas. "And also to use it as a place to figure out how to adapt. We'll be an experiment, and other people can see it and learn."

The Atlantic hurricane season peaks in August and September because that's when the water is warmest. All else being equal, rising ocean temperatures mean fiercer winds and crueler rains. The relationship between global warming and hurricanes is one of the most hotly debated subjects in the field of climate change. Storm activity moves in cycles that can take decades to run. Computer model outputs are often vague. And the data is fragmentary: advancing technology—storm-chasing aircraft and weather satellites—means we know a lot more about the recent past than we do about what happened even a few decades ago.

Nonetheless, the science is coalescing around the conclusion that while the warming seas may not generate more storms, they have the power to transform tropical squalls into full-blown hurricanes, to pump major storms into regional cataclysm. For those forced to suffer through them, the distinction between frequency and intensity is likely to seem academic. By the end of the century, the carbon dioxide in the atmosphere could triple the number of hurricanes reaching category five on the Saffir-Simpson Hurricane Scale, with winds capable of ripping apart all but the strongest reinforced concrete buildings. Regardless of climate change, hurricane experts are expecting a surge in the number of storms, the reemergence of a long-term cycle that last peaked in the 1960s. If this natural rise in frequency is coupled with warming waters, the upcoming decades could feature the beginnings of unprecedented meteorological violence.

Between 1966 and 2003, only one major hurricane reached the coast of Florida. In 2004 alone, a record-breaking four made landfall. Two of them pummeled the same stretch of coast within three weeks of each other. A fifth swiped at North Carolina.

The following 2005 season was the most active on record, with fourteen hurricanes. An unprecedented four storms reached category five status. Hurricane Emily was the earliest category five storm ever recorded. Hurricane Wilma, off the coast of Cuba, was the most intense Atlantic hurricane ever measured. Florida and Louisiana were hit twice. Texas was struck once. Hurricane Katrina cut across southern Florida as a category one storm, gorged itself to category five on the unusually warm waters of the Gulf, and spun into Louisiana and Mississippi as a category four with winds of roughly 140 miles per hour and a thirty-foot surge of water. It shredded the levees in New Orleans, flooded 80 percent of the city, killed 1,800 people, and stranded tens of thousands more on rooftops and in makeshift shelters, surrounded by a fetid mix of sewage, industrial chemicals, and lake water. It caused $81 billion in damages, making it the costliest disaster in the history of the United States.

As the Florida finale to the two violent seasons, Hurricane Wilma made landfall just north of the Keys. The storm had weakened as it ripped across Mexico's Yucatán Peninsula, but then spun across the Gulf, feeding on its warmth. Wilma passed seventy miles northwest of Key

West, dealing the island a glancing blow as it crashed into southern Florida. The town had been evacuated five times in fifteen months, but it had never taken a direct hit. Two seasons of near misses had left residents fed up and broke. This time, very few had evacuated. For a while, it seemed the island had dodged the season's final bullet. It hadn't.

Billy Wardlow, then the Key West fire chief, was at his shorefront headquarters when he saw the water begin to lift the boats in the docks across the highway. The storm surge was as gentle as they come, but the winds chopped at the waves. "It came just straight across the street, right up the apron on our fire station," said Wardlow. "At the car lot next door, you could see the water reaching the car windows. The electric windows started going up and down because of them shorting out. We had the police and fire boats parked in the little alley. The water got so deep it floated them up off the trailers." The waters rose six and a half feet before slowing. Breakers rolled through the city streets. Palm trees thrashed in grayish waters. "All I could think of was Katrina," said Wardlow. "What am I going to do with these people? We have twenty-five thousand people, and we have water coming, and we can't get the trucks out of the fire station." Then the winds subsided, the waves calmed, and the ocean slowly sunk away.

"Wilma could have been much worse," said Matt Strahan, the meteorologist in charge of the National Weather Service bureau in Key West. "The wind hit here with peaks of eighty or ninety miles per hour. It shoved the water into the shallows of the bay, and it had nowhere to go so it came down across the Keys. By the time the surge was up at its peak here, the storm was passing. So then the backside winds came out of the north and helped push it across us. It was a pretty gentle surge, though. It flooded everything, but it didn't knock down anything."

Strahan had ridden out the storm in his newly built hurricane-resistant headquarters, and when I visited he offered to show me around. The floors of the $5.1 million building sat nearly fourteen feet above sea level, higher than the theoretical maximum surge. There were rooms with cots and couches, a standby generator, and racks of car batteries to make sure the power never goes off. The windows had shutters and hurricane glass. The walls were constructed to withstand sustained winds of 165 mph. "It's basically built like a fort," said Strahan. The bathrooms formed the sanctuary. Enclosed in reinforced walls, they were rated to withstand winds of 255 mph. The heavy, bolted metal door opened inwards, ensuring it would not be blocked by debris. "The designer claims it can take a car," said Strahan. "You get two-hundred-mile-per-hour winds, and you get cars airborne. You want to know what the experts think can happen in Key West? This is it."

Strahan pulled up a map of the town on his computer. His offices sat on White Street, the dividing line between what the locals call Old Town and New Town. To the southwest lay the historic district, the higher ground of original settlement. The other side had been developed in the last few decades during the long lull in hurricane activity. Most of it was four feet above sea level or lower. Almost all of it had flooded during the Wilma surge. Strahan indicated a particularly low-lying section of Key West named Stock Island that was undeveloped in 1950. About a third of it was given over to the town's golf club. The rest was spread with trailer parks, commercial docklands, and cheap housing. "Since this was all built, they've not had a major hurricane hit this area," said Strahan. "Just the fact that they've now put people where they didn't used to live means they're exposed to danger."

A former stretch of swamp and marsh, Stock Island was separated from the rest of Key West in 1846 when a brutal hurricane broke over the island, destroying all but six of the town's six hundred houses and blowing a channel through the marsh. In 1919, a near miss of a category four storm swept Stock Island with fourteen-foot waves and a ten-foot storm surge.

"Obviously, these disasters don't happen very often or nobody would be living there today," Strahan continued. "The chances of one happening are relatively slim, but it's going to happen someday. There are seven thousand one hundred people who live on that island. Even if half of them evacuate, you take thirty-five hundred people and wash them away, and you've exceeded the death toll of any hurricane in history except for Galveston in 1900. On just one little island in the Keys that nobody lived on fifty years ago."

"People in Key West will tell you, 'My house has been here for over a hundred years, so it must be a strong house because it's seen a lot of hurricanes,'" said Strahan. "I researched the hurricane history this year. It turns out that in Key West records go back to 1870. And since 1870 Key West has never experienced one hundred and ten miles per hour or stronger sustained winds. Maybe the houses are well built, but there's no historical record to suggest that they're right. I know a person who lives in a trailer on

Stock Island whose floor must be five feet above sea level. And he tells me, 'My trailer's been there fifty years. It must be a strong trailer.'"

A storm does not have to make landfall to cause lasting damage. Hurricanes look monstrous on a satellite feed, and a city under direct assault undergoes a nightmare. But the strongest winds hug up against the hurricane's core. All but the communities closest to the center can hunker through the lesser gusts of even the greatest of tempests. The churning waves of a storm surge have the broader impact, but the most widespread harm has more to do with evacuations and insurance than wind or water. A hurricane endangers a city, but a string of storms threatens the economy of an entire region.

"Even if we're not hit, it causes us to lose a week or two weeks of business," said Ed Swift, a Key West developer. "We have a law that says that during a hurricane evacuation, we actually close the hotels and force the customers out forty-eight hours before the storm hits. Of course, nobody in the hurricane business wants to make the wrong call. So they make the early call, and they evacuate your business."

"And when that happens, all your guys and gals that depend on tips—all your bartenders, all your waiters and waitresses—they're out of business," he said. "Those folks are living usually month to month. They can't pay their rents. The landlord, maybe he's got three apartments, he's like 'Shit, I can't pay my mortgage.' So he puts off fixing the roof, and the roofer guy doesn't do his work."

"The community is good at responding," he said. "But

what those hurricanes did when they came like that—bang, bang, bang, July, August, bang, September, boom—is they wore the town out. It has taken all this time since the flooding to rebuild the spirit."

After Katrina, insurers cut back on exposure all along the coast. From the Texas docklands to the beaches of Cape Cod, coverage suddenly became much harder to find and much more expensive. Allstate, the nation's second-largest insurer of cars and homes, had suffered $5.7 billion in catastrophic losses in 2005. Of the ten most expensive catastrophes it had ever paid out, seven had come in the previous two years. The company had already begun cutting back in Florida. It now announced it would not be renewing policies in Texas and Louisiana. In 1938, a category three hurricane had crashed across New England, killing six hundred people. Studies suggested that a repeat could result in $200 billion in damages. Allstate announced it would not be writing new coverage for homeowners in New Jersey, Connecticut, Delaware, New York City, or Long Island. "We believe what the scientists are telling us," a spokesman told *Newsweek*. "The country is in the beginning of a period where we've sustained more frequent and more intense hurricanes. We believe it would be bad business to continue to add to our risk."

The 2004 and 2005 seasons had set the insurance industry reeling, generating 5.6 million claims and insurance payouts of $81 billion. By comparison, losses from hurricanes during the previous two years were $2.2 billion. "There was a guy in Zurich that you could call up the day after a major hurricane who could pretty accurately tell you what the losses for the company would be," recalled Chris Walker, who at the time headed the Greenhouse Gas Risk Solution Unit for Swiss Re, the world's largest reinsurer. "He could look through the systems and see what we've insured in that region, see where the storm had gone, and pretty much guesstimate."

"Where he was wrong this time is he didn't factor in business interruption," said Walker. "The sheer magnitude of the devastation meant that the McDonald's that was damaged, for instance, if everything else had been equal, the windows and roof could have been fixed within days. And it would have been operational. Now it had this extended inability to operate. Even if it was fixed, there were no customers. There were no sources for food."

"The 2004 situation was four hurricanes in rapid succession hitting the same area," said Chris Winans, vice president for media relations at the American International Group, the country's largest insurance company. "So you had a new risk factor being built in. It's one thing to say how a hurricane damages a structure. It's another thing to talk about how the second one right behind it redamages it. At what stage are you at repairing from the first?"

To those in the industry who had been sweating the impact of global warming, the twin seasons seemed to confirm their worst fears. As early as 1992, the investor Warren Buffett had warned that "catastrophe insurers can't simply extrapolate past experience." "If there is truly 'global warming,'" he wrote, "the odds would shift, since tiny changes in atmospheric conditions can produce momentous changes in weather patterns."

Weather-related losses had been accelerating at a rate of growth ten times faster than premiums and the overall economy. From roughly $1 billion a year in 1970, they were reaching an average of $17 billion a year in 2003. Swiss Re keeps three climatologists on staff. In a report released just before the 2004 season, the company had predicted that another decade of global warming would hit insurers with between $30 billion and $40 billion in weather-related claims every year. And with Katrina, the industry was facing that from hurricanes alone. Had the future arrived already?

"The whole underpinning of what insurance is about is that the past is an accurate predictor of the future," said Walker, now the North American director of the Climate Group, a coalition of businesses and government that works for action on global warming. "And if it's no longer an accurate predictor—because climate has changed—then you're in trouble. You don't know what a good price for housing insurance in Key West is anymore. Is a one-in-a-hundred-year event now one in fifty? One in twenty? One in ten?

"If it was a linear movement—one degree rise will equal one more storm—you could calculate and probably it would be an actuarial dream," said Walker. "But it's not. In 2007, Florida had a major fire during the rainy season in every single county simultaneously. That's pretty weird. I don't think of Florida as being a dry place subject to forest fires. That's out west. That's California. I always say 'climate change' as opposed to 'global warming.' Because it's going to be warmer some places. It's going to be cooler some places. Some places will have more rain. Some less.

If climate change changes the data that you're relying on as an insurer, then how do you price? How do you model? And if you don't price and model, are you only gambling?"

Droughts in Darfur and hurricanes along the Gulf Coast were all potential symptoms of a change sweeping the entire world, said Robert Muir-Wood, the chief research officer at RMS, a leading modeling firm: "There are some areas that appear to be completely immune at present, but it's likely random which areas have been hit and which areas haven't." The rising rates all along the coast were simply a recognition of the increased risk. "Our position is that we're responding to the fact that activity is higher," he said. "It's not that climate change is out there in the future. It's already happening. And the future will probably look a bit like this."

The world has not yet put a formal price on carbon. Drivers don't pay at the pump for the expected impact of their commute. Nor do we factor the future damage from running air conditioners, lighting, and kitchen appliances into our electricity bills. Yet global warming does have a cost, one that for the moment is monetized only in higher prices for certain types of insurance. If rising rates near the water reflect a world made riskier at least in part by climate change, then communities like Key West and New Orleans, and all those up and down the Atlantic and Gulf coasts, have begun paying for the world's emissions.

Wilma's waters were quick to recede, but the damage from the storm seasons had only begun. The town's residents were still cleaning up from the flooding when the state provider of wind insurance announced the repeated landfalls had drained its reserves and put it $1.5 billion in the red. Rates in Key West, it said, were underpriced and would be rising. Premiums, already among the highest in the state, suddenly as much as tripled. "Insurers are ruining our economy with their predictions," said the Reverend Raymond Shockley, the pastor at the Church of God. "They're using fear to destroy us. We've weathered hurricanes for a long time and survived them. But we cannot survive wind insurance." Shockley's church, a few blocks down White Street from the National Weather Service, endured fifty thousand dollars in damages from Wilma's winds and waters. Flooding at the parsonage tore down a fence and infiltrated the ground floor, ruining the carpets, wrecking the appliances, and pulping the pressed-wood furniture. Gusts of wind blew through an old-style louvered window and streaked the back wall of the sanctuary with rainwater. "We had a kind of expressionistic crucifix for a while," Shockley said. "And we repainted and repaired."

Shockley is a big man, with a broad, pink face and a nose like a pale plum over a thin, white mustache. He wore a black, short-sleeved, button-up shirt and rimless eyeglasses. His white hair was carefully combed. There was deep honey in his accent and a touch of theater in his gestures. The church he runs is a family affair. His son, a bishop, teaches Bible classes and leads the music during services. His daughter-in-law works the Sunday school and helps in the church's Christian bookstore, the only one in a hundred and fifty miles. The congregation supports orphanages and churches in Africa and Central America. In Key West, it provides a place of gathering and prayer for the town's residents and a place of refuge for its visitors.

"If the Bible had been written in America, then Jesus would have said the prodigal son went to Key West and wasted his substance," said Shockley. "We're the far country. That's what people do here. They come here and get immersed in the lifestyle and begin to party and drink and do drugs. Pretty soon they expose themselves to some sexual adventures that make them sick."

"Then they discover the church," he said. "More than once we've had a prodigal son come in here, or a daughter, come to themselves, and find they need God and a direction for their life. I've bought more than one bus ticket to help a guy leave here. I've had to go to the county jail and counsel with bridegrooms who got married here and before their honeymoon could be over were busted for drugs and their bride left standing on the street."

We were joined by the church's youth pastor, a twenty-six-year-old computer technician for the Monroe County school system named Andrew Hish. He wore sneakers, faded blue jeans, and a dark blue collared T-shirt. His hair was cut short and boyish.

"Andy's in the process of trying to buy a house," said Shockley.

"There's a lot of houses on the market," said Hish. "It's just finding one where the insurance rates aren't killing you," he said. To buy a home, Hish needed a mortgage, for which the bank required he have wind, flood, and homeowners coverage. The premiums on top of his monthly payments put just about the entire island out of his range.

"Andy has a nice wife and two children," said Shockley. "There's a lot of pressure on him. He could make more money somewhere else, where it costs a lot less to live."

"We've lost several teachers," said Hish. "Very good, young teachers because they just can't afford to live down here because of this. A lot of them would be willing to stay if they could get their own place. But it's just the wind insurance they are paying. I know teachers who are married, that pull in easy one hundred thousand dollars a year, and they can barely afford it."

"And each time one of those families leaves, they're leaving a church too," said Shockley. "Churches are mostly supported by discretionary funds. So when you take and double and even triple the wind insurance on my church people, they have less money to share with the church." Shockley lifted himself from behind his desk. Choir music had begun to play from the sanctuary. A prayer meeting was due to begin. "Then, to top that off, when I came here our total year's cost of insurance was about fifteen hundred dollars," he said. "Today I have to raise fifteen hundred dollars a month."

After Wilma, the fire department lost thirteen of its seventy-three employees. Some, like Wardlow, the former fire chief, retired. But most moved away. "They just couldn't afford the insurance rates on top of the mortgage and everything else," said Wardlow. Key West's cost of living had already made finding and retaining workers a challenge. Now homeowners were considering selling. "It literally is a battle for the soul of this community," one resident told the *Miami Herald*. "A business can raise prices and pick up an extra couple hundred dollars, or a thousand, ten thousand, depending on its size," said Swift, the developer. "It's a lot harder for a guy working for a wage when his landlord comes and says, 'Geez, I'm really sorry, but they doubled my insurance, and I've got to charge you an extra grand this year.'" "Your workforce are the ones who are making thirty or forty thousand," said Wardlow. "And that's what you need here. You need police. You need firemen. You need people in the restaurants. You need people in the motels."

TWO YEARS AFTER KATRINA, NEW ORLEANS WAS STILL FAR FROM recovery. The city had returned to two-thirds of its prestorm population, but police still operated out of trailers. Less than half the city's schools had reopened. Ten of the city's twenty hospitals remained closed. From the interstate, I looked out on abandoned cinemas and gutted big box stores. Weeds grew in parking lots, and derelict shopping malls hoisted wind-damaged signs. In the worst-hit areas, house after house was a boarded, gutted shell. Stagnant waters had transformed thriving middle-class neighborhoods into blocks of urban blight.

The long-term economic effects of the catastrophe were as corrosive and more lingering than the floodwaters themselves. New burdens exacerbated the Big Easy's already famous corruption, crime, and bureaucracy. Property taxes had jumped as the city shifted its weight on a shrunken citizenry. Insurance rates had soared. The once comfortable were often at the margins. Those at the margins had been pushed out. When I visited, the weather was hot and muggy: the feeling of thunderstorms about to burst. The city was still struggling to breathe.

"Insurance truly is going to make or break the economic redevelopment," said Paige Rosato, an insurance lawyer. "You have people who still, two years after the storm, are just trying to get back into their homes. You have people wanting to sell their homes; they have buyers, but they can't get insurance for it, and so the sale doesn't go through." At the time of the storm, Rosato was a researcher in a law firm specializing in defending insurance claims. Her life was transformed by Katrina. Rosato and her family had taken refuge from the winds at her parents' house in Baton Rouge, and when they returned, they found their home had been completely destroyed. "One neighbor's tree went diagonally through five rooms of the house," she said. "Another neighbor's tree cut off the front. Those five rooms were totally exposed to the wind and the rain. It was like a stack of cards that just imploded on itself."

Until the storm, Rosato's job had largely consisted of defending insurance companies' clients from third-party lawsuits. But as the waters dried and New Orleans began to pick itself up, it became clear the firm would now be protecting the carriers from aggrieved policyholders. Rosato decided to quit. "In researching the insurance code and defenses, I began to understand that the same defenses that apply to the business commercial claims were going to apply to the homeowners' claims," she said. "And I just didn't want to be a part of that. I wanted to be a part of making that not happen." She stayed in Baton Rouge, began lobbying for insurance reform, and took on clients who felt wronged by their insurers.

Rosato wore a turquoise silk shirt with cloth buttons and matching brown pants with turquoise flowers. Her sunglasses were pushed back on her head. A silvered angel hung from her pearl necklace. As we talked, there were times her eyes welled with tears. "At the time, my in-laws were ninety years old. They had their insurance policy for over sixty years with the same company and were initially given a pittance offer. They had sustained four and a half feet of water in their home. They had been relocated in Texas."

"In March, they had been saying, 'We want to come back; we want to see; we want to see,'" Rosato said. "My brother-in-law flew them back. We picked them up at the airport. We drove out to the property. There were FEMA trailers lined up all along their street. My father-in-law—we were there no longer than ten minutes—says, 'I've seen what I needed to see.' And he died ten days later. He knew he was not coming back."

Deductibles, low payouts, undercoverage, and the rising cost of construction meant many of Rosato's clients who had owned their homes for decades now had heavy loans. Insurance payouts had not covered the cost of rebuilding. "They haven't had a mortgage in twenty years," she said. "They now have one, and in addition to that they're being faced with a whopping insurance bill that many times is equal to their outstanding mortgage payments. And it's devastating people. I've had so many clients, elderly, and not just one or two, sit in my office crying, wondering how they're going to make it. It used to be, Are you going to pay for your medicine or are you going to pay for your food? Now it's, Or are you going to pay for your insurance? It truly is the roof over their head."

Meanwhile, in the lightly regulated commercial market, insurers were cutting exposure wherever they could. "I've seen rates quadruple," said Anderson Baker, president of Gillis, Ellis and Baker, a New Orleans insurance agency. "I've seen wind deductibles increase to the point where the buyer of the insurance policy says, 'Forget it. I'm never going to have windstorm damage of that magnitude. I'll just go naked.'" Ninety percent of his clients' buildings had lost their coverage after the storm. Not only had rates jumped; insurance companies were hesitant even to write the policies. Baker had flown to Bermuda, London, and Washington trying to find someone who would. "We never used to have to buy wind separately from the other perils," he said. "Now it's not unusual for us to have at least two policies, and then depending on the magnitude of risk, we may be layering three, four, five, six companies to get to the level of coverage that's necessary. Prior to Katrina, I don't think I had ever layered a policy in my career. Now it's commonplace."

A New Orleans native, Baker wore a seersucker suit striped in white and blue. He had a creased but youthful face and thick hair and wore a red tie and shiny black shoes. He fretted over what insurance prices were doing to his city. Rates had stabilized at between two and three times their preflooding levels. Meanwhile, elsewhere in the country, away from the coasts, they had begun to drop. "There's a lot of people who are saying, 'That's one more reason I'm not going to stay here,'" said Baker. "And they begin to make the decision to follow everyone to Atlanta or Dallas or Houston. That's the real problem that I see with insurance prices. It's not that it's expensive everywhere. But that it's expensive here."

Rising rates added pressure on local businesses, who could neither easily pull up and leave nor compensate by boosting their prices. "The local office supply store owner who is only paying insurance based on his local shop can't very easily raise the price of his paper," said Baker. "Because he's got Office Depot to deal with. They buy on such a worldwide basis that New Orleans is a rounding error in their insurance scheme. And then people begin to make decisions about where they're going to buy their paper. They say, 'You know what? If my homeowners premium is so much more expensive, I'm going to cut my expenses any way I can.' And one way is buying paper where it's cheaper. So it puts the poor local guy kind of in an awkward spot."

Baker had just returned from Washington, where he had lent his voice to a proposal to extend federal flood insurance to cover wind damage as well. It was an idea the insurance industry had opposed, and I asked him why, with his earnings based on commissions, he was so worried about rising rates. "Arguably, they're good for me personally," he answered. "But they're lousy for this city. I'd be happy to have our rates back down by a third if we have insurance that lets this economy recover." He wasn't optimistic. "It may be five years before this market straightens itself out from an insurance perspective," he said. "And that's assuming no storms. If we have another storm within a hundred miles of here, it's 'Katy, bar the door.' Last time I evacuated, I took a rucksack. Next time I might take a steamer trunk."

Fifty-four percent of Americans live within fifty miles of a coast. Increasingly, they will be feeling the squeeze of climate change. Before Katrina, the most expensive storm in history was Hurricane Andrew, which slammed into Florida south of Miami in 1992, killing sixty-five people and causing $26.5 billion in damages. Since then, the population of the state has grown 30 percent. According to one study, if the same storm were to hit today, losses would reach $55 billion. "People quite literally don't care how many hurricanes hit the state or how dangerous it is," said Robert Hartwig, chief economist at Insurance Information Institute, an industry association. "So strong is the lure of the sea, apparently, that people discount and are willing to ignore the risks of coastal living."

But the tide is turning. People are able to disregard storm dangers as long as they're abstract. Insurance premiums make them concrete. In 2006, more people moved out of Florida than into it. The population of the Keys has dropped by 6 percent since the last census. Between 2005 and 2006, when local lobbying brought a reprieve from the rising premiums, nearly eight thousand people fled Key West.

Yet coastal insurance may, if anything, be artificially low. Even with the higher premiums, there's a shortage of insurance companies willing to take the risk. The majority of home and business owners in places like Key West and New Orleans are covered by government-run carriers. With rates set as much by political as actuarial concerns, the result is huge exposure, falling ultimately on the taxpayer. The 2005 storm season pushed the National Flood Insurance Program $20 billion into the red. In 2006, Florida state officials kept their fingers crossed and hoped for the best. The state insurer had written more than $400 billion in policies, and a major storm would have overwhelmed the plan. A catastrophic one could have bankrupted the entire state. "Ultimately an insurer's rates have to reflect the risk," said Hartwig. "If they don't, the insurer cannot operate. He couldn't make good on his obligations."

As the globe warms, rising rates will accelerate the depopulation of places like Key West. Residents will find themselves in a losing battle against the cost of living. Vacation homes, which already make up a large proportion of the town's houses and apartments, will multiply as the remaining middle class gradually gives way to those who can afford the insurance. The transformation will start out slowly. Those who have paid off their mortgages may choose to go without coverage. But when a big storm hits, those who find it impossible to rebuild will give way to those who can afford to write off the risk.

All along the coast, climate change will transform a way of life. "I think this will lead to development coming in the form of high-rise condos instead of individual homes," said Becky Mowbray, the insurance reporter for the *Times-Picayune* in New Orleans. "While tall steel and concrete buildings are more intrusive than small, low-slung homes, condos could be good from an environmental perspective, because it could turn more land back to nature. However, I can also imagine that it could turn beachfront vacationing into more of a class/luxury issue than it has been in the past, since only the wealthy could afford to buy at the beach. My bet is that the mountains become the new retirement destinations of choice because of the rising cost of insurance and development at the coast—I think the Florida retirement dream is over."

• • •

After Katrina, some in New Orleans questioned the wisdom of rebuilding a coastal city so prone to devastating floods. The idea was too controversial to gain any political support, but the poststorm economics are having a greater impact than any government policy could have. Recovery has been slower in the city center, but the higher ground across Lake Pontchartrain has boomed. The largely white suburbs of St. Tammany Parish have been the only part of the hurricane-hit region to gain population since the storm. "I don't think the city of New Orleans will in my lifetime get back to much over three hundred thousand people," Ivan Miestchovich, the director of the Center for Economic Development at the University of New Orleans, told National Public Radio. "The downtown is going to be convention oriented, tourist-and-visitor oriented. It will still have a role to play in financial services and banking. But that said, most of those will have some kind of a satellite arrangement somewhere else."

"If rising insurance rates will help turn cool Key West into more of a second-home resort playground for the rich, what happens when the same dynamics are unleashed on one of the poorest metropolitan areas in the country?" said Mowbray. "Part of what made New Orleans such an authentic tourism destination was that it was affordable, and real musicians could afford to live there. What happens if they can't return? And while the wealthy may be willing to pay whatever it costs for a water view in Key West, you can't run a city like New Orleans, which stands on tourism, oil, and shipping, without workers."

"I believe these higher rates are here to stay, which will

push development inland as living in New Orleans or elsewhere on the coast becomes less affordable," she said.
"That puts centuries of investment in New Orleans at risk,
and I believe will create a more car-dependent commuter
culture as the density of the city is zapped and housing and
development spread between [the nearby suburban cities
of] Baton Rouge and Covington."

Writing in *Science* magazine just before Katrina, Evan Mills,
an environmental scientist at the United States Department of Energy's Lawrence Berkeley National Laboratory,
sketched out a scenario for the future of the insurance
industry under the changing climate. He predicted that
property and business-interruption losses would continue
to force rates up, while extreme temperatures, worsening
water quality, and vector-borne diseases would add new
costs to health and life insurance. Meanwhile, major insurers would begin to see lawsuits against their clients as victims of global warming turn their attention to the emitters
of greenhouse gases. There would be more and more years
when the industry isn't profitable.

Nor will the problem necessarily be confined to the
North American coasts. In addition to their astonishing
destructiveness, the 2004 and 2005 seasons held several
more surprises. In 2004, a storm that meteorologists would
later nickname Catarina struck with category two strength
at the southern Brazilian coast, ripping at roofs and wash-
ing the streets with waves. The storm damaged forty thousand buildings and devastated the rice, corn, and banana
harvest. The reason it hadn't initially been given a name
was that nobody was watching for it. Never in recorded

history had a hurricane-strength storm been spotted in the South Atlantic.

The next year, late in the season, meteorologists watched in wonder as Hurricane Vince formed in the far east Atlantic and spun itself out against the coast of Spain. Damage was minimal, but it was the first time a tropical cyclone had hit the Iberian Peninsula. It might not be the last. Some models suggest that global warming could set off hurricanes in the Mediterranean, putting at risk some of the world's most densely developed coastlines.

Next door to where I was staying in New Orleans, a man was packing his house into a pickup truck. John Nelson was tall and slender, with a long, thin nose on a long, thin face. His hair was blond, pulled back into a ponytail, thin but not thinning. He had a bit of a southern drawl, offset by precise enunciation. His house was half packed. Boxes lay piled in the living room. A gray carpet was rolled up on the polished wooden floor. Framed paintings still hung on walls. We sat at a dining room table spread with magazines and bills, and he explained why he was moving. Clearly upbeat by nature, he was struggling to stay that way.

Two years earlier Nelson and his family had been flooded from their homes by Katrina. This time, they were being pushed out by insurance. "There's extreme wealth in this town, and there's extreme poverty," said Nelson. "The middle class is just barely treading water." He and his wife owned three small houses in a row in a dicey part of town. The central building, a rental under construction, blew over in the storm. Next door, his wife's nearly completed studio took six feet of water. Their home, a raised cottage, was flooded by a foot. "It did the same amount of damage as eight feet," said Nelson. "All the electrical had to be redone. All the plumbing had to be redone. We had to gut the whole thing and start the whole thing again." Nelson had been a collector of rare books, with first editions of Mark Twain's *The Adventures of Tom Sawyer* and *Adventures of Huckleberry Finn* and an 1860 printing of *Don Quixote*. Large and heavy, they had been kept on bottom shelves, where they were ruined.

The insurance paid out, but it paid out slow, and while Nelson and his wife rebuilt, they bought the house in which we were sitting. The previous owner had lent it to a contractor, who had trashed it, and the asking price was low. "Our monthly payments—we figured it out before we moved in—would have been two thousand dollars, which is pretty inexpensive," said Nelson. "And then with insurance and everything else, we were thinking a crazy outside figure would be twenty-five hundred dollars."

The first blow was the property tax. The house was reassessed for what they had just paid for it, then reassessed again. With a concurrent jump in his insurance rates, the monthly payments had reached thirty-six hundred, "a tremendous amount of money just to start off with before you eat," said Nelson. "I figured our property taxes would go up. They would have to. We've lost so much of the city. I'm not opposed to it. But I really thought we'd be able to get a handle on the insurance rates. They would go down, and it would make it more reasonable to live here." Instead, just over a year after they moved in, they received yet another letter from their carrier. Their premium would rise again. "I called my Realtor that day," he said. "And he said, 'I've gotten eight or nine phone calls in the last two days saying the same thing.'"

The family would be staying in New Orleans, on the second floor of a building they had bought to work in. "I could have started working at Wendy's somewhere else and probably would be having just as much money as I do right now," Nelson said. "I've got less than when I was eighteen years old, as far as spending money. And I'm in the hole." Nelson and his family were moving to a worse part of town, to an area that had been heavily flooded. They were trading covered porches and tree-lined streets for boarded-up buildings and layers of peeling paint. "I feel like I'm having to evacuate again," Nelson said.

3

"A SPECTACULAR BIT OF GROWTH AS TIMES GET HARD"
EUROPE, MIGRATION, AND POLITICAL BACKLASH

The island of Lampedusa is famous among Italian tourists for its giant sea turtles, which is strange, considering so few of them nest in its beaches. It's not a big island, and its only town is even smaller; I could walk from the airport to my hotel. Even though it was after ten o'clock, late enough to darken even a Mediterranean summer evening, the center buzzed with tourists. Young couples walked hand in hand. Parents pushed their baby carriages. The air smelled of cooking smoke and of salt. In the central square near a modern church, a man on a synthesizer played an accordion waltz. Gray-haired men danced their dyed-hair wives in slow circles across a polished marble compass rose. A bougainvillea's lavender flowers trembled in the breeze.

Lampedusa's central street, a brick lane called Via Roma, doesn't follow the water's edge, but from the way the darkness cut low between the houses, I had the impression I was surrounded by the sea. Stores offered coral and turquoise jewelry, beachware, sandals, inflatable rafts, lotions, hats, sequined shoes, cotton dresses, and tur-tle souvenirs: ironed onto T-shirts or fired out of clay, big or small, beaded or plain, stylized or naturalistic, glazed or naturally rough and pink.

Several decades ago, when Italy was home to ten thousand turtle nesting sites, Lampedusa played an insignificant role in their breeding. The island is a speck far off Sicily's southern coast, a stretch of crumbling limestone less than six miles long, closer to Tripoli than the Italian mainland. Its rocky coasts offered few places where a turtle could dig to bury her eggs, just short stretches of paper-white sand embracing a tiny cove.

But for the turtles, the important thing was the island's isolation. As Europe's Mediterranean coast bloomed with hotels, houses, and beach resorts, Lampedusa—impoverished and half forgotten—remained undeveloped. "A turtle doesn't come to shore where there's light, or noise, or movement," said Daniela Freggi, who runs a turtle rescue and research center for the World Wide Fund for Nature. Loggerhead turtles can take up to forty years to mature and mate, meaning that many would-be mothers are just now returning to the built-over nesting grounds of their infancy. "If her beach is not available, she will stay in the water," said Freggi. "And the eggs just spill out, and they die. They become food for the fish." Italy now has fewer than twenty nesting sites, including the one in Lampedusa. "This means less genetic variability, fewer possibilities, less of a future generation," said Freggi.

Most of the three hundred to four hundred turtles that pass through Freggi's center each year are rescued from trawler's nets or longlines. In many, thick iron hooks lodge in their digestive tracts, so the center runs a surgery—complete with X-ray machines, ultrasound, anesthesia, and oxygen—to pull them out. When tangled fishing line cuts off circulation to a flipper, the limb gets amputated. If Michael White, the center's chief scientist, is right, the project's patients are destined to become fewer. White, a retired British naval engineer who used his pension to work his way to a Ph.D., argues that climate change is doing to the sea what developers have done to the coast: making it tougher for turtles to live.

Long lives mean long reproductive cycles, the Darwinian equivalent of a heavy shell and slippery flippers, and it's unlikely that the turtles will be able to adapt to sudden spikes in temperature. "When I was working in Greece, I was studying nesting beaches," White said. "What I was finding was in the hot years, you were getting a large number of solidified eggs. They would be hard-boiled just by the heat of the sand." Even less extreme temperatures can be a danger. A hatchling's sex is determined by the heat in which it develops. Nest temperatures greater than about eighty-four degrees Fahrenheit yield more females. Cooler sands produce more males. "You can look around the world, and largely what we're finding now is that many of the nesting beaches are above this pivotal temperature," said White. "All you're getting produced in many places is females. And so you've got a second route to extinction, by running out of males."

Turtles may have put Lampedusa on the tourist maps, but the island has become increasingly associated with another phenomenon, one that will increase with climate change: illegal immigration. The island is geologically African, a crumb of oceanic crust from the eastern edge of the underwater Tunisian plateau buckled like the Alps by the collision of continental plates. Tucked into a corner between Tunisia and Libya, Lampedusa is politically a sliver of Europe in African waters, and it has become the continent's most visible entry point for those desperate for a seat at the global table. Offshore sightings of half-sunken boats overfilled with immigrants have become so common that Italian television treats them with a sense of uncomfortable routine, almost as if reporting on spates of unpleasant weather.

Climate science is better at spelling out trends than making specific predictions. We may not be able to say if a given stretch of beach in Greece will become too hot for turtles, but we can predict that temperatures across the Mediterranean will rise and that animals adapted to cooler climes are likely to suffer. Likewise, when it comes to the disruption of human habitats, the poor of the world will be disproportionately hit. Most developing countries lie in the tropics or in deserts, where rising temperatures and climatic disruption will be hardest to absorb.

Richer countries will be better positioned to weather the changes, through technology, through increased spending, or simply through the good fortune of being farther north. Drought-stricken Australia is spending billions on solar- and wind-powered desalinization plants. The Dutch are preparing for rising sea levels with houses that can float through a flood. Canadian, Russian, and Scandinavian farmers could even benefit from wetter, milder winters. The result is likely to be a further widening of global disparity, an increase in immigration pressures in border regions such as Lampedusa, and political reverberations across the developed world as politicians and the public struggle over the proper place for the less fortunate.

In 2007, nearly twenty thousand would-be immigrants arrived in Italy by sea, the bulk of them in Lampedusa. Not all of them made it. The week before I arrived, the Italian military pulled fourteen bodies from the sea sixty miles off the island's coast. Two weeks later, the Tunisian Coast Guard hauled in another twenty corpses. That year, at least 471 people who attempted the crossing were reported dead or missing. The boats usually launch from Libya. A passenger—with luck one who has piloted a boat before—is given a compass and told to steer north. It's a dangerous, uncomfortable trip. Most who take it can't swim, and the smugglers pack their charges so tightly that it's impossible to move. The boats are old, chosen to be abandoned, and the 170-mile crossing can take days. When doctors treat the immigrants on their arrival, the most common ailments are dehydration, sunstroke, poisoning from engine fumes, and painful cramps from spending days in the same position.

In 2003, *Time* magazine interviewed a young Somali named Abdi Salan Mohammed Hassan who spent eight months traveling from Mogadishu to Tripoli, including a ten-day drive across the Sahara in the back of an overheated flatbed truck. He then paid eight hundred dollars for a cramped spot with eighty-five other immigrants on a small fishing boat, which broke down in the open sea: "The passengers are withering in the sun, their exposed skin blistering and peeling; at night . . . their skin shivers and crawls. By day they see ships passing in the far distance. Some men try to make oars out of pieces of wood. The boatman burns shirts for smoke signals. All in vain. On the fifth day, another ship appears, and the boatman dives into the sea and disappears, the first among them to die. Others are soon to follow. Driven half-mad by thirst, people begin to drink seawater and moan with hideous cramps from the salt in their bellies. Some lean over the side to scoop the sea into their mouths, fall in and drown. . . . By day thirteen, more than forty have been dumped into the sea, and two dozen people are sprawled across the floor of the boat, barely breathing. 'I saw people dying all around me,' Abdi Salan recalls. 'I was just waiting to die, too.'" When the Italian Coast Guard boarded his ship, they found thirteen bodies and fifteen survivors.

Those who study immigration break its driving forces into pull-and-push factors. Europe and the United States pull at the poor with the promise of work, higher living standards, and security. In Africa, the Middle East, and parts of Asia, overcrowding, unemployment, war, and natural disasters cut at their feet and push them away. Immigrants like Abdi Salan know they are gambling their lives when they set out for Lampedusa. But for the tens of thousands for whom hope and hardship overwhelm the risk, climate change will only drive them harder.

In a report headed by British rear admiral Chris Parry, the United Kingdom's military strategists predicted that the number of people living outside their country of origin will grow. Driven by "environmental degradation, the intensification of agriculture, and pace of urbanization," the world's expatriate population will expand from 175 million now to 230 million in 2050. In the United States, eleven retired admirals and generals, writing for the

Virginia-based national-security think tank CNA Corporation, predicted that migration would be climate change's greatest challenge for the country: "Already, a large volume of south to north migration in the Americas is straining some states and is the subject of national debate. The migration is now largely driven by economics and political instability. The rate of immigration from Mexico to the United States is likely to rise because the water situation in Mexico is already marginal and could worsen with less rainfall and more droughts. Increases in weather disasters, such as hurricanes elsewhere, will also stimulate migrations to the United States."

The Red Cross says that environmental disasters already displace more people than war. Christian Aid, a London-based charity set up in 1945 to deal with the mass displacements of the Second World War, estimates that there are about 163 million forcibly displaced people in the world. Between now and 2050, the charity predicts, another 250 million people will flee floods, droughts, famines, and hurricanes caused by climate change. Another 50 million will lose their homes to natural disasters, some of them fueled by climate change. Fifty million more will seek refuge from extreme human rights abuses and conflicts. In some of these, as in Darfur, climate change will again have played a role.

"There's no question that, in many parts of the world, climate change—changes in sea level, desertification, deforestation—are having an impact on migration flows," said Brunson McKinley, director general of the International Organization for Migration. He had come to Lampedusa to see the immigrants' first point of arrival, a reception camp run by the Italian government. We spoke at the airport as he waited to board his plane. "Within countries, people leave unproductive farmland to go to the big cities and look for jobs," he continued. "Very often those people don't find the jobs they're looking for in the big cities, and they think about going farther, going on to where the jobs are."

"I don't know whether you subscribe to the notion that even some of these cataclysmic events are related to climate change," he said. "I'm not a scientist, but I would argue that there may be a connection. The melting of the ice caps, for instance, does cause a rebalancing of the tectonic plates, and when they shift, you have these cataclysmic events, earthquakes, tsunamis, volcanic eruptions, mudslides, and of course they have an important impact on the population. Many people lose their livelihoods. Are there more hurricanes, typhoons, and tropical storms than there used to be? Some people think so. They often have the result of ruining plantations, ruining farms that were productive and self-sustaining, and forcing the people who live there to go elsewhere. We've seen that happen in Central America and lots of other places. It's not science fiction; it's present-day reality."

Two years earlier, McKinley would have been unable to visit Lampedusa's detention program. It was run as an Italian military zone, and journalists and international observers were forbidden entry. Then an Italian journalist named Fabrizio Gatti threw himself into the waters offshore and waited to be brought in. In his description of the week he spent posing as a Kurdish asylum seeker, Gatti describes a camp that is overcrowded, dirty, and degrading, an improvised fenced-in enclosure buffeted by the backdraft of tourist-laden Airbuses. The camp is designed to sleep 190 detainees. During his stay, the population reaches 1,250. Immigrants sleep on sidewalks or under bunk beds, Muslims are mocked for their faith, and beatings are commonplace. On Gatti's first full day, lined up for roll call, he finds himself straddling a rivulet of spill-off from the toilets. The guards force him to sit nonetheless. Four days later, he watches as a group of uniformed carabinieri and a man in street clothes welcome a new group of immigrants:

"'Strip,' says [the man in street clothes] to a kid in a tank top who is trembling from cold and from fear. He doesn't understand. He doesn't move for a full minute. 'What is the problem?' yells the carabiniere, and he slaps him on the head. The immigrant, pale and skinny as a skeleton, trembles. Another slap. All the people standing nude in front of the carabinieri get slapped. For half an hour the carabinieri have been talking of making a corridor. . . . They soon show what it means: a string of six foreigners headed for the camp passes between them and each takes his ration of blows. Four carabinieri deal out four blows each. Finally, the sergeant . . . appears. But he doesn't reproach anyone. 'Is this guy giving you trouble?' he asks the man in street clothes, and he lands a punch on the sternum of the skinny immigrant, who doesn't understand what he's done wrong and who is still standing, immobile, in his tank top."

In the outcry that followed Gatti's article in the Italian newsweekly *L'espresso*, the International Organization for Migration, the Red Cross, and the United Nations' refugee agency were given a permanent presence to monitor the camp and advise immigrants of their rights. Detainees would stay for no longer than a day or two before being transferred to other facilities. Management of the camp was licensed out to another company. On the day I visited, the barbed wire was limited to a single strand running across the top of the green entrance gate. A boatload of twenty-three Somalis had just arrived and waited outdoors to be photographed and fingerprinted. The women stretched out on a concrete bench, scarves pulled over their heads. The men were teasing one another, laughing like mountain climbers who had followed a perilous path and finally reached their peak. Several wore shiny gold sneakers given to them by the center.

I wasn't allowed to speak with the detainees, but my guide, a young woman named Paula Silvino, the vice director of the camp, was at pains to show me how much it had improved. She brought me to the dormitories, long prefabricated buildings holding forty beds, and to the mess hall where a group of West Africans were watching soccer in English on a satellite-fed widescreen television. But she lingered longest in the kitchen. "We give them three meals a day, trying to respect the habits of our guest," she said. "We offer fish, rice, vegetables, no meat. We have products that aren't discount, products of high quality. And we have a machine that seals the plates one at a time, and it maintains the freshness, so there aren't big, military-style pots. This way it's hygienic."

In addition to shoes and a change of clothes, each arrival is given a phone card and ten cigarettes a day. The center also provides the coast guard with kits of water, fruit juice, and crackers for immigrants just off the boat, and tries to stock newspapers in Arabic, as well as a soccer ball. "Yesterday, I went to play with them myself," said Federico Miragliatta, the camp's director. "There's one that's been playing nonstop since he arrived. We had a boat arrive at five thirty in the morning. This guy was playing ball at six."

I was offered a different perspective on the camp and its occupants when I accompanied the deputy mayor of Lampedusa on a trip to the center of the island where workers were putting the finishing touch on a new reception center. Angela Maraventano picked me up outside city hall, a three-story building with a flaking facade. She wore tight jeans and white sandals with silvery sparkles. A black sleeveless shirt was pulled tight over her heavy breasts, and throughout the day she was rarely without a cigarette in her mouth. Her plastic purple-framed sunglasses spent most of the time pushed up to the top of her head.

Maraventano, a forty-three-year-old mother of two, is a member of Italy's Northern League, a far-right political party that has at times advocated for secession of the country's northern regions, and she shares her party's hard stance on illegal immigration. When the Northern League's minister Umberto Bossi flamboyantly declared in 2003 that the immigrants' boats should be fired upon, Maraventano demurred: "I don't exactly think we should shoot directly at them," she said. "Shoot over their heads, maybe." There are nearly no immigrants in Lampedusa outside the camp— once free there would be nowhere to go—but Maraventano has been able to tap into a deep anger at what's seen as spending on illegal immigrants at the expense of the local community. Recently elected, she told me she received between two hundred and three hundred calls a day, and indeed she spent most of the day on her Nokia telephone. Her demeanor was loud and gleefully aggressive, a far-right Erin Brockovich.

Her car had broken down that morning, so she was driving a borrowed four-door hatchback, gray and covered with a thin layer of white dust. In the passenger seat, a woman in heavy makeup and a sleeveless shirt with a Lampedusa turtle motif introduced herself as Vincenza Filippi, the vice prefect for immigration at the Ministry of the Interior. After stopping for gas (the attendant pointed out his Northern League flag, the southernmost in all of Italy), we drove out of town into a moonscape of white rocks and desert scrub.

The new center was much bigger than the one by the airport. Large two-story buildings of metal frames and insulated panels sat on either side of a wide concrete pad. We had stopped to pick up a prominent Northern League parliamentarian, Angelo Alessandri, and Maraventano directed her commentary to him. Lampedusa's only secondary school is crumbling; the flooring is chipped, the plaster has peeled away from its rusted steel supports, and in at least one room, metal scaffolding holds up the ceiling. As we toured the new reception center, Maraventano gave a running description—half joking, half furious—of how she would occupy the $12 million compound and turn it into classrooms. Blond and mustachioed with a tuft of hair beneath his lower lip, Alessandri wore a green, loose-fitting linen suit. He twirled his sunglasses in his hand as he listened.

The center was planned for a capacity of five hundred immigrants and could be expanded to hold a thousand. Each building had four rooms on either side of a shower area. Each room had its own air conditioner and smoke detector and six beds, each of which could be extended into a bunk.

"Hey, Angela," said Alessandri. "How many kids would fit here?"

"You just need to move the panels," she said. "You have to move the panels, because for classes the rooms are small, no? So you move the panels, and make bigger rooms."

The beds, they were told, would be bolted to the floor. A fence would string the perimeter, and infrared beams would alert a central control room of any breaches.

"Who causes the most trouble?" asked Alessandri.

Filippi answered, "Usually the North Africans are the most"—she searched for the word—"exuberant."

"Those are the ones we should send home," he said.

Alessandri had brought his girlfriend, and as the party walked towards the mess hall, she stopped.

"Imagine the suffering that they've gone through, just to reach here," she said.

"But when they get here, what do they do?" said Maraventano.

"The kids end up begging at stoplights," said Alessandri. "The women become prostitutes. To me, it's racist just to let them in. You can help them in their own country, and it costs you a lot less. You can build a village for ten thousand people with the money we've spent."

His girlfriend was unconvinced. "But once they're here, once they've arrived, what can you do?" she asked.

"I repeat, the solution is not to make them come," he answered.

"This is the mess hall for our kids," said Maraventano. "Finally the kids of Lampedusa have a mess hall."

"Nice," said Alessandri. "Nice, if it's for the kids."

"No kidding," said Maraventano. "Stainless steel, walk-in fridge, everything."

"Just think," said Alessandri. "If we held them in the old center, and kept them there for one month or two, and we made them understand that they'd spend six months in the old one, maybe they wouldn't come."

"That's right," said Maraventano.

"Here, we say, 'Good morning, have a rest, and then you're free,'" said Alessandri. "That's the mistake."

"Did you see?" Maraventano asked as we drove away. "Lampedusa can't accept what this government is doing. This little handkerchief of an island, these five thousand people, we can't submit to this kind of phenomenon. It's shameful. We're five thousand, and they're building for another thousand? It's a village inside a village. Are we joking?"

Immigration politics swing between two polar emotions, fear and empathy. Dread of the foreign—of cultural clashes, crime, and competition—can coexist with sympathy for the downtrodden, even in the same electorate, often in the same person. It's the politician's challenge to walk the rope between the two. It's nearly impossible to deter the arrival of people who have risked the Sahara, Sudanese border guards, and the Mediterranean, and the public has shown it won't stand for blatant mistreatment.

The result in Italy is a schizophrenic policy, in which the government neither welcomes nor refuses the immigrants. Italy doesn't release figures on the fate of those who pass through its detention center, but United Nations High Commissioner for Refugees (UNHCR) says that one in eight of those who land at Lampedusa are eventually granted asylum. The rest are mostly funneled through a series of detention camps, held, questioned, and eventually given a sheet of pink paper that informs them they must leave Italy within five days. And then they are free. Many head north to friends and family in Rome or Milan. Others pass on to other countries.

As climate change will mean a rise in immigration, it must also mean a rise in immigration politics. Pressure will build on the politicians of the developed world to come down on one side. The success in Italy's southern-most town of a political party that stands for northern superiority can be chalked up to the dominance of fear over empathy. The Northern League tapped Lampedusa's resentment towards those who wash up on its shores and won a seat in the local government, a platform to highlight the issue nationwide. Its politicians understand the power of the image; a human tide of dusky masses crashing on Italian shores will tip the scales towards fear.

Later that summer, I headed to another European island to see how another anti-immigrant party was trying to do away with the scales altogether. The London streets were darkened by a rain that nobody seemed to notice. The July weather flashed from rain to breaths of sunshine cut short by sudden downpours. My first meeting was in the eastern suburb of Barking and Dagenham, a predominantly working-class neighborhood where the xenophobic British National Party (BNP) has won twelve of the borough's fifty-one council seats. Having been told that Barking was the London stronghold of a party "committed to stemming and reversing the tide of nonwhite immigration," I was surprised by how racially mixed the neighborhood was. Black women pushed their babies in strollers. A group of young South Asians walked by with tennis rackets in their hands. An ashen European passed with a newspaper folded under his arm. Schoolgirls ran by in burgundy uniforms and white head scarves.

When I arrived a couple minutes late for my appointment at a pub behind city hall, Richard Barnbrook, the BNP's lead councillor, was sitting at a picnic table outside, holding his papers against the gusting wind. The son of a Buckingham Palace guard, Barnbrook is one of the BNP's cleanest faces, forty-six years old, a teacher and an artist, unstained by the party's thuggish roots. He wore a brown suit and a brown tie over a gray shirt. His face had the pale, wind-burned look of someone who spends a lot of time under the clouds. A displaced tooth in his lower jaw left a tiny gap, and when I spoke he fixed me with his blue eyes and tightened his mouth seriously. As the party's most prominent London politician, he was planning runs for the mayor's office and for Parliament. I quickly lost track of the number of cigarettes he smoked. London had just instituted an indoor smoking ban, and we sat outside until the wind turned to rain.

The big issue in his constituency, he told me, was housing, specifically the city's plan to build new homes in the area. "There's no good land left," he said. "All that's left is the contaminated lands or the marshes. My ward is called Gorsebrook. The next one is Mayesbrook. Another one is Eastbrook. A brook is a river. The floodplains are there for one good reason; they flood. And to build a house on that is total nonsense."

"The southeast of England has a major water shortage," he said. "Be it global warming, be it environmental change, the water shortage in south England is really quite drastic. And yet the biggest housing projects are happening in the southeast." Barnbrook told me he had belonged to the Labour Party in the days before Tony Blair, but joined the BNP in 1999 when mass immigration into his part of London resulted in "the almost soul-searching loss of the community around my town." I asked Barnbrook if what he was really speaking about was not the environment but immigration. "The link between the two is quite simple," he said. "More people coming into the borough means that the green land goes and is replaced by cement." By talking about floodplains and water shortages, Barnbrook was replacing the politically unpalatable—if openly stated—fear of foreigners with a more acceptable empathy for the environment.

The BNP is the British analogue to Italy's Northern League or Jean-Marie Le Pen's National Front in France; it's part of a trend in the growth of anti-immigrant parties as the European public and its politicians fret about fitting Islam into their Christian and secular institutions. The BNP hasn't had its continental counterparts' electoral success, and that has forced it to think a little differently. The

United Kingdom's system of government doesn't allow for much proportional representation, so unlike in Italy, where a vote for the far right works towards putting another member in Parliament, a ballot cast for the BNP can usually just as usefully be thrown away. But as British concerns about immigration have grown, the party has started to make small gains. During the 2006 local elections, it more than doubled its council seats, from twenty to forty-six. Most of its support came from areas like Barking and Dagenham, working-class neighborhoods with large numbers of recent immigrants. In the elections for national government it chooses to contest, the BNP usually comes in fourth, outpolling the Greens and other small parties.

The party's success is due in no small part to the efforts of its leader, Nick Griffin, a longtime right-wing politician who since taking over the BNP in 1999 has concentrated on softening the party's harder lines. Shaved heads and bomber jackets have given way to blazers and ties. Marches and violent confrontations are avoided. A policy of forced repatriation of immigrants was replaced with one in which legal arrivals would be paid if they left. When Ian Cobain, a reporter for the left-wing *Guardian* newspaper, joined the group in 2006 in an attempt "to take a glimpse behind Griffin's facade of normality," he found that the transformation extended even behind closed doors: "Its activists often shun such words as 'black' or 'white,' even when talking at party meetings. Many of its activists have accepted the need, in Mr Griffin's words, to 'clean up our act, put the boots away and put on suits.'"

"I heard phrases . . . uttered by BNP members many times and, after several months, came to understand their precise, nuanced meanings," Cobain wrote. "'Nice areas' I quickly understood to signify predominantly white areas. 'Quiet areas' are places where black and minority ethnic people live, but keep a low profile, and don't compete too hard for jobs, school places or sexual partners. 'Troublesome areas' are places where black people do just the opposite. 'No-go areas' are places where black and minority ethnic people are in a majority."

"In my seven months as a party member I heard very few racist epithets, and no anti-Semitic comments," he wrote. "Such language appears almost to be frowned upon in Griffin's post-makeover BNP."

Nick Griffin's conversion seems to have come during a 1998 trial in which he was found guilty for inciting racial hatred with articles that called the Holocaust a "Holohoax." "Take my experience of revisionism," he told a conference in New Orleans organized in 2005 by David Duke, a former grand wizard of the Klu Klux Klan. "I was convinced at that stage that it's so damned obvious, the flaws in certain arguments, that all you have to do is put them in front of someone and like you they'll be converted. And what made me realize I was wrong was seeing the jury. Twelve good men and true—a couple of them ethnics, there you go, it was a liberal sort of area—and when we were discussing politics, they were actually alert, they were really interested." But when he began speaking about the Holocaust, he lost them: "Within thirty seconds, literally, I saw their eyes glaze over."

"It actually goes beyond that," he concluded. "You can't even build an organization . . . that can become successful in politics. You simply can't do it talking about those issues that interest you unless they interest the people."

While the BNP remains centered around immigration politics, Griffin has struggled to broaden the subjects with which it is associated. Where possible, its politicians have adopted language that, as Barnbrook showed in Barking, attempts to replace fear with empathy. Rather than cry out against an influx of Asians and Africans into government-subsidized housing, the BNP takes a different tack: "Everybody knows the problem there, but they're nervous about agreeing with you for fear of being called racist," Griffin said. "So if you say: what the Labour Party is doing is racist; it's breaking up the local community; it's discriminating against local people on the grounds of their color. Then all the conditioning that's been told to everyone who has been through school or watched television in the last forty years—that racism is bad—suddenly switches on to pushing them towards our conclusion."

"There's nothing unusual about it," he said. "It's standard politics. It's just that nationalists in most countries in the West for some decades haven't had the faintest interest in playing standard politics. They want to have their own little perfect club. It's what Blair did with the Labour Party, what Thatcher did with the Conservative Party. It's what all political leaders and groups that are serious about changing the society do to set about getting power. They package their core message in something that is widely popular."

. . .

I met Lee Barnes, the man most responsible for the BNP's blend of anti-immigrant environmentalism, in the town of Chatham. Located to London's east, it lies just over a broad hill from Gravesend, the port on the Thames where Joseph Conrad opens his *Heart of Darkness* on a gloomy ship, describing England from the point of view of an ancient Roman colonist: "The very end of the world, a sea the color of lead, a sky the color of smoke . . . Sandbanks, marshes, forests, savages—precious little to eat fit for a civilized man, nothing but Thames water to drink."

In the cash-strapped BNP, Barnes, an unemployed graduate of legal studies, handles court cases. Unofficially, he acts as a one-man think tank, working on what he calls "strategic ideological development," positioning the party for the future. Barnes wore a faded black T-shirt over his thickened frame and blue jeans torn along the seam at the cuff so they split over his sneakers. He had a handlebar mustache and a tuft of hair that hung off the bottom of his lip. His long hair had pulled away from his temples. Around his neck hung a Thor's hammer, a symbol of his Pagan religion. We sat in a café off of Chatham's central walk, while Barnes, who refuses to eat factory-farmed meat, ate a shrimp and mayonnaise sandwich, peeling the crusts off thick slices of white bread. I asked the occasional question, but mostly I just let him talk.

The brand of British nationalism Barnes described harked back less to the glorious days of the British Empire than towards Conrad's forgotten colony, where Romans shivered and cursed the locals and the weather. "Modern nationalism says I am not superior to you," he said. "How can I be superior to a Zulu who's sat on his land and taken care of his land for thousands of years? If you want a world where the environment is protected, it's about local people protecting local cultures and traditions. It's not about supremacy. It's about removing the supremacy. We believe we're a special people. But if I were somebody living in the rain forest, I'd believe I was special. And if I was an African Zulu, I'd believe I was special. And if I was an Inuit in Greenland, I'd believe I was special."

"Nationalism and environmentalism were linked at the moment the first human being set foot in his country," he said. "You had the ancient Celts, the ancient druids. Their temples were the forests. There was a connection between the people and the land. It was the Roman Empire that sundered that connection. Then you had Christianity that came in and did exactly the same, with its idea that man has dominion over the earth. And then Economics came in and said that Economics is more important than the environment. All of these things have to drop down a level, and the environment has to come up a level."

The association of environmentalism exclusively with the left is a modern phenomenon. Jorian Jenks, the editor of the British *Mother Earth* magazine, which pioneered organic farming and foreshadowed the green movement, was a senior member of the British Union of Fascists before World War II. According to the historian Jonathan Olsen, the Nazis themselves had a strong green wing: "Almost immediately after taking power the National Socialist regime undertook far-reaching environmental legislation, including reforestation programs, wetland protection, laws limiting industrial development, and aesthetic measures, such as the attempt to design the German Autobahn according to environmentally sensitive principles." In the United States, the 1920s conservationist Madison Grant, who saved many species from extinction, was a noted eugenicist and the author of *The Passing of the Great Race*, which argued that Nordic Europeans had achieved racial superiority by adapting to the harsh, northern climate. "Environmentalism has been hijacked by the left," Barnes said. "It's no longer green; it's red. Those traitors. I despise the green movement. 'Oh, we oppose nuclear energy.' Well you can't oppose climate change through carbon dioxide emissions and then oppose nuclear energy. 'Oh, we support mass immigration.' How can you support mass immigration when you're supposedly an environmentalist party?"

"The environment is a big issue coming," said Barnes. "It's not such a big issue now because the impact hasn't been felt yet." Neither Barnes nor Griffin accept that increased levels of carbon dioxide drives climate change, but both believe environmental pressures will lead to hard times ahead, be it through the spawning of new diseases due to globalization or the world simply running out of oil. "At some point, the way we're abusing and exploiting the environment will trigger an environmental crisis," said Barnes. "The point is to start preparing now by putting the ideology and the language of environmentalism into nationalism. It's the crisis of the future. That and overpopulation, which are symptoms of exactly the same thing. People that have an irreverence for the land breed like rabbits."

Barnes's views find surprising echoes among the environmental establishment, where they dovetail with arguments that the world's population is unsustainable. James Lovelock, the former NASA scientist who originated the Gaia theory treating the Earth as a living organism, has long fretted that the planet will be unable to support its current population. The World Wide Fund for Nature has warned that people are consuming natural resources 20 percent faster than nature can renew them. The Sierra Club has undergone repeated takeover attempts by a close-the-borders wing, which has tried to add immigration reduction into the environmental group's purview.

The organization that draws the most explicit links is the Optimum Population Trust, a British environmental think tank whose most prominent supporters include the chimpanzee researcher Jane Goodall and Paul Ehrlich, a professor at Stanford and the author of *The Population Bomb*. The group estimates that the United Kingdom's population needs to be cut by half to thirty million on environmental grounds and that the earth is unlikely to be able to sustain more than three billion people, less than half the current figure. It has called for limiting British families to two children, describing not having a third child as "probably the most effective action people can take to halt climate change" and calculating in a press release that each new UK citizen releases over his lifetime an amount of carbon dioxide equivalent to 620 return flights across the Atlantic.

A significant impediment to the group's goals is immigration, which the Optimum Population Trust says accounts for 66 percent of the United Kingdom's population growth between 2001 and 2005, compared with 2 percent in the 1950s. To reach a sustainable population, the group argues, immigration must be capped at a level that matches emigration. "We're calling for net-zero or neutral immigration in this country, and that would allow for quite big flows in and out," said Rosamund McDougall, who sits on the group's advisory council. "We're just asking that it be balanced. Because we're very, very densely populated, and our footprint and carrying capacity are just way over the top."

"There is an internal division building up in the Green Party in Britain, which is obviously generally very left wing," Nick Griffin told me when I met him in an east London pub on another rainy afternoon. We sat by a corner window and watched the weather darken the sidewalks and the roads. "There's now a section of them who are starting to say you can't talk about saving the environment in Britain without some kind of immigration control. In due course, if we have the luxury of a bit more time on our hands, we will look at ways to get in there and widen the split and hopefully peel some of those people off that movement and attach them to us."

"We consider ourselves the only logical green party in Britain," Griffin said. "There's nothing logical with wanting to save the whales and a small species of ants and at the same time wanting to encourage or put up with mass movements of population, which is contributing to the biggest wipeouts in human ethnic and cultural diversity that the world has ever seen. Every person we take from the third world with a tiny climate footprint and bring them into the Western world, we're massively increasing their impact of carbon release into the world's atmosphere. There's no doubt about it, the Western way of life is not sustainable. So what on Earth is the point of turning more people into Westerners?" Rising immigration pressures and worries about climate change, said Griffin, will drive the environmentalists and the nationalists together, like two trains approaching the same platform from opposite directions. "You end up at a parallel position," he said. "It's fairly easy to jump."

But while the cocktail of climate change and immigration may green the xenophobes and drive some environmentalists to the right, few liberal greens are likely to join the BNP itself. The coalitions are more likely to take place within the mainstream conservative parties, where the two movements may find themselves to be convenient bedfellows. What the Germans call Watermelon Coalitions (green on the outside, red on the inside) could yield to Camouflage Coalitions (green on the outside, brown on the inside) as both sides begin to borrow the other's arguments against what they increasingly see as a common threat.

The BNP's views on the environment are unusual for a nationalist party only for the extent to which the party embraces them. During the last French presidential campaign, the far-right candidate Jean-Marie Le Pen played down his xenophobia in favor of a new theme: environmentalism. In Austria, a new party led by the right-wing populist Jörg Haider promotes strict immigration controls, support for organic farming, and a "green" tax to take into account the full environmental cost of a good or service.

In the United Kingdom, where the far right remains a minor political force, the net effect is likely to be a slide to the right among the major parties as formerly radical ideas gain currency. But, in places like northern Italy, anti-immigrant groups already enjoy considerable cachet, and the left and right compete in a delicate balance. Shifting coalitions could have real electoral impact. In Belgium, where the Flemish nationalist Vlaams Belang has become one of the country's largest parties, a skillful expropriation of environmentalism could yield seats in government.

Even if the far right proves unable to turn warming worries into xenophobic fear, governments will increasingly come under pressure to be seen as doing something about the waves of climate-shattered poor sweeping up on their shores. In 2007, the United Nations Office for the Coordination of Humanitarian Affairs announced it had responded to a record number of droughts, floods, and storms. Of the thirteen natural disasters it had been called on to assist with, only one—an earthquake in Peru—was not related to the climate. In the previous record-holding year, 2005, the office responded to ten disasters, only half of which had anything to do with the weather. As this trend continues, illegal immigration will become a perennial issue, and politicians are more likely to find electoral benefit in splashy displays of shutting the door than in far-sighted efforts to address the root causes. For years, France ran immigration camps at the entrance to the channel tunnel to England. Italy has pressured Libya to detain immigrants on its side of the Mediterranean. While efforts like these are unlikely to have much of an impact on immigration, they are likely to proliferate nonetheless as politicians try to send a message to voters and immigrants alike: this is what happens when you enter illegally.

Meanwhile, immigrants will likely have messages of

their own, an appeal for the empathy of the electorate. The showdowns may echo what happened in northwestern Australia earlier this decade, when the government tried to deter arriving immigrants with a publicity campaign aimed at prospective arrivals. "You will NOT be welcome," read posters distributed across the Middle East. "You WILL be kept in detention centers thousands of kilometers from Sydney and you could LOSE all your money and be sent back." Australia's camps scattered across the outback ended up filled with thousands of immigrants. In 2000, more than 150 asylum seekers declared a hunger strike. At least a dozen sewed their lips together to protest the length of their detention. Following public outrage, the government closed the worst of the camps in 2003.

Pressure will mount on the countries from which immigrants are fleeing to either take back their residents or to stop the flow. But while some may try to comply, these efforts are also unlikely to have much impact. In cases where civil war or environmental catastrophes spark spikes in arrivals, we might expect to see military and humanitarian interventions similar to United States president Bill Clinton's 1994 foray into Haiti, when rising numbers of boat people following a coup on the Caribbean island convinced him to mobilize twenty thousand American troops to reinstate the democratically elected president, Jean-Bertrand Aristide. When the military junta backed down at the last minute, the invasion became a humanitarian intervention.

Immigration is a diverse and dispersed phenomenon. Migration flows may follow broad patterns, but for each major entry point, driving factor, or desired destination there are thousands of individual paths, motives, and objectives. Unless developing countries radically change their character, there's likely little they can do to stem the flow. In many areas, ethnic diversity on the scale of New York's or London's will likely become the rule rather than the exception. Multihued, multilingual streets will become the norm.

Whether host countries are injured or enriched by their swelling foreign ranks will largely depend on how well immigrants are assimilated. Rear Admiral Chris Parry, the British military strategist, has warned of "reverse colonization" as the Internet and cheap flights allow migrants to stay connected to their homelands. "The diaspora issue is one of my biggest current concerns," he told a conference of military experts. "Globalization makes assimilation seem redundant and old-fashioned. . . . Groups of people are self-contained, going back and forth between their countries, exploiting sophisticated networks and using instant communication on phones and the internet."

Parry's worries are no doubt exaggerated, but he's not the only one seeing the future in improbable but worrisome terms. In a provocative article in the *International Journal of Environment and Sustainable Development*, Peter Wells, a professor at Cardiff University, paints a picture of an emerging green/right alliance, triggered in his scenario by growing frustrations over democracy's inability to deal with global warming: "A modern Green Junta is unlikely to arrive with tanks on the streets and the overnight capturing of control. Rather, it creeps upon us through multiple small steps—each one justified by 'necessity.' It combines the national interest with a contemporary version of making the trains run on time, it highlights the external amorphous dangers, it makes dissent equivalent to being unpatriotic and antisocial. It clings to the tokenistic manifestations of democratic society while removing their substantive content."

While I was speaking with Griffin, I made a few notes about his appearance, described the creases under his eyes, the hook of his nose, then scratched them out. He looked unexceptional, a politician's blandness alleviated by a pudginess that fleshed around his eyes and softened his jawline. He wore a blue shirt over suit pants. His hair was cut short and parted. We had eaten as we talked, and he lingered over his pint of beer before inviting me to join him on a drive to Canvey Island, a depressed stretch of urban sprawl built on marshland at the mouth of the Thames. The BNP was inaugurating a new chapter, and Griffin had scheduled a visit.

His car was a Ford Mondeo station wagon with darkened windows driven by his bodyguard, a bodybuilder with the chest and head of a professional wrestler: bald, broad, with a thin goatee and a brow that wrinkled over his eyes. After a few false turns, and a stop for a quick dinner and so Griffin could change his shirt, we arrived at a strip of dilapidated nightclubs and casinos facing a raised embankment protecting the buildings from the Thames. A wintry wind blew a froth across the waves, which tossed, tipped, and sucked at the light. We entered a door next to the Las Vegas Casino and walked up a worn, carpeted staircase with mirrors on both sides. My first impression was of a rundown strip club, but I was still surprised to see a sign that warned: "Strictly No Physical Contact Between the Customer and the Performer."

Griffin is aware that an anti-immigration platform, even if it were to sweep up all the green votes, is an unlikely election winner, but he has been crisscrossing the country nonetheless. He and Barnes have convinced themselves—and much of their party—that oil shortages, mounting debt, and mass immigration will combine to pull the United Kingdom, and indeed the rest of the world, into a crisis. The subsequent turmoil, if not civil war, will offer them an opportunity. "Obviously there's only so much we can do in normal times," said Griffin. "But you have to establish a degree of political legitimacy, if not acceptability, before you get hard times. And as they come, then all sorts of things that haven't been politically possible up until then, they become possible."

"When the economy crashes, the main people to be hit will be the middle class," Barnes had told me. "And they will embrace radicalism, and you'll see a new form of political agitation based on protecting their lifestyles. You'll have a popular movement demanding social change. You'll see us getting more and more of a political mandate. At the same time the state, the corporations, the media will become more and more repressive. You'll get more and more people marching on the streets. And then there will be a confrontation at some point." I asked Griffin if he agreed with this vision of a Ukraine-style orange revolution. "Apart from violent revolutions, there are very few examples of a ruling oligarchy with a particular closed mind-set giving way," he said. "Other than by mass protest. And them getting the feeling that unless they give way gracefully, they'll end up swinging from lampposts. Which is what all orange revolutions are about. It's not about wearing orange jackets. It's about saying there's so many people here, you can't shoot us all, so pack your bags and go."

The strip club, it turned out, had been deconsecrated, and was now a nightclub specializing in Elvis and Tom Jones impersonators. We walked into a meeting in full swing, with about fifty attendees seated in chairs on the dance floor. On a stage that once featured a dancer's pole, three men sat at a table draped with the Union Jack. English flags hung against the windows and on the wall. There was the strong smell of stale beer, and the bottom of my feet stuck to the floor. After a couple of speakers had their moments, it was Griffin's turn. He had given a speech the night before, at a neighboring chapter, so tonight he would be taking questions. "It's one way of highlighting that we're different from the other parties," he said. "Tony Blair. When did you last see him surrounded by real people? He surrounds himself by real people's children because they don't ask awkward questions." The Griffin in front of the crowd was a different man than the one with whom I'd had two rather lazy meals and a wandering, almost disinterested conversation. Instead of standing behind the tables as the others had, he stood on the edge of the stage on a platform that brought him closer to his listeners. "We're the party that knocks on people's doors," he said. "You can see that we have no spin here, so fire away." With his suit jacket on, his shoulders seemed squarer. He gestured constantly with his hands. He brought them together, pointed towards the audience, held one clenched in front of him like a fist. He answered the questions in the staccato rhetorical style of Tony Blair facing the opposition. He took question after question and looked, frankly, parliamentarian.

On Iraq: "We believe that our men and women need to come back immediately, and the very first platoon that comes back needs to be stationed at this end of the channel tunnel."

On deporting immigrants: "The government already rounds up a lot of illegal immigrants, so they can't complain when the BNP rounds up all the illegals."

On white flight: "If that had been done to any other people on this planet, it would be called genocide. Because genocide is not just about machine gunning, gassing people. It's also destroying their community."

On the future of the BNP: "The other parties will be forced to coalesce against us. That's a fantastic position to be in when times are hard. Because the other parties have already told the British people that you're the different one. And so that's where we'll be in five years' time: poised for a spectacular bit of growth as times get hard."

4

"AT A NEW FRONTIER"
BRAZIL, UNSETTLED ECOSYSTEMS, AND DISEASE

The city of Manaus is a metropolis of nearly two million people carved into the center of the Brazilian rain forest. It lies just upstream from where the black currents of the Rio Negro merge with the milky waters of the Rio Solimões and form the Amazon River. Perfectly placed for the exploitation of the rubber trees once found exclusively in the South American forest, Manaus rode its monopoly through the Industrial Revolution and into the twentieth century. Between 1850 and 1920, it grew ten times in size, blossoming from sleepy backwater into the "Paris of the Jungle." It was the first city in Brazil to install trolleys, the second to put up electric streetlights, and the first to be endowed with a university. Its cultural centerpiece, the Teatro Amazonas, a Renaissance-style opera house completed in 1896, boasts French tile work, Italian crystal, and British cast iron.

The city's boom was busted when an agent for London's Royal Botanic Gardens smuggled seventy thousand rubber seeds out of the country, allowing the English to undercut the South American harvest with plantations in

Malaysia, Sri Lanka, and tropical Africa. But Manaus clung to its preeminence as a geographic, political, and cultural hub and regained its ascendancy in 1967 when Brazil turned the river port into a free-trade zone. Nokia, Sony, Kawasaki, Harley-Davidson, Honda, and Kodak set up shop in the industrial district, and the Paris of the Jungle became an Amazonian Hong Kong. Even now, prices for electronic goods in the city run roughly a third cheaper than in the rest of the country. The city assembles more than half of Brazil's televisions. Foreign tourists may know Manaus as a launching pad for jungle excursions, but for Brazilians it's a place where you arrive with empty suitcases and leave carrying extra freight.

"Manaus is a big town isolated in the middle of the forest," said Flávio Luizão, a coordinator for the Large Scale Biosphere-Atmosphere Experiment in Amazonia, a Brazilian-led international initiative to study the jungle. "But it has quite a lot of pollution. Manaus is a quite rich town because of the industrial district, so there are many cars, and the traffic is very bad. Then, almost all our power is from thermo electric: burning oil. The dam we have nearby was a complete ecological disaster. A huge lake, but too flat, it produces only enough, and maybe not even, for the industrial district. All the houses and shops in Manaus receive their electrical power from burning oil." The result is a plume of traffic exhaust and power plant fumes rising above the green of the jungle like smoke from a snuffed-out candle.

Yet the image of the city as an island of smog in a sea of clean air masks the true picture of pollution in Brazil. Every dry season, swaths of the Amazon fall to the chain saw. As trees are felled and set ablaze all across the forest, and particularly in the south, smoke from thousands of fires white-out the sky. The sun becomes a faint spot struggling to show through. Airports across the region are forced to shut down. Rising particles of black carbon spur the formation of towering clouds that refuse to dissolve into rain. Compared with these areas of heavy deforestation, Manaus looks clean. "There is a place in the southern fringe of the Amazon where we have towers and equipment collecting data for more than twelve years," said Luizão. "You can see the huge peaks of the aerosols in the atmosphere during the dry season." During the worst of the burning, the air quality compares badly even with that of central São Paulo, Brazil's famously smoggy commercial capital.

Nor do the problems stop when the fires go out. The newly exposed soil releases nitrogen dioxide, a pollutant most commonly associated with cars. It floats into the lower atmosphere where sunlight converts it to ozone, a toxic gas that attacks lung tissue, aggravates asthma, and stunts plant growth. Brazilian researchers have found that during the dry season, ozone levels triple, nitrogen dioxide concentrations are ten times greater, and the amount of carbon monoxide in the air jumps by factors of fifty to ninety. "Most of the pollution problems in Brazil are from deforestation and not from urban pollution," said Luizão.

When I visited Brazil, the rainy season had just begun. The fires had long since burnt out, and the air was clear. I flew first to Manaus and then to Porto Velho, the capital of Rondônia, a small Brazilian state wedged up against the border with Bolivia, and then I drove south. Deforestation in the Amazon begins with a road. The government builds a highway, and loggers, legal and otherwise, cut offshoots to pull out the most valuable lumber. On satellite images of Rondônia from 1996, the penetration appears as a fish-bone pattern: the state is a patch of green, lightly scored by highways with a cross-hatching of secondary roads. In pictures taken ten years later, the fish has fleshed out. Farms and ranches fill in the arteries where the forest once stood. Outside the areas of Rondônia where cutting is prohibited—conservation areas and indigenous reserves—the forest is completely fragmented. Small stands of trees separate the fields. Nearly a third of the land in the state has been cleared.

The road from Porto Velho was flanked with pastures. Pale green hills rolled past the occasional dark cluster of what once had been impenetrable forest. I was driving with Eraldo Matricardi, a scientist at the state environmental protection agency who had returned to Brazil a few months earlier, after completing his doctorate in Michigan. There was nothing to show that the area we were passing through had recently been jungle. "My father is a farmer," said Matricardi. "He sees the piece of forest that he left behind as a problem, an obstacle. This is the idea of everybody who came to the state: 'I came to cut down forest.'"

Rondônia began losing its jungle in the 1970s when Brazil's military government began encouraging the poor to move from the overcrowded southern portion of the country. To the top brass it was an elegant solution. It exploited untapped resources, filled in a military void, and eased pressure for land reform. For the settlers, it was a chance for a new beginning. Only the trees stood between them and the rich land beneath. I looked out the window as Matricardi drove. Fences ran along the side of the road. Thick-shouldered white cows grazed in the fields. It looked like Idaho, if Idaho had palm trees. "Deforestation is good," said Matricardi. "At least, that's how it seems to the local community."

It wasn't long, however, before Brazil's colonization project ran into an unexpected challenge. The arrival of the settlers coincided with an unprecedented outbreak of malaria, a disease caused by a mosquito-borne parasite that leads to fevers, nausea, chills, and aches. In 1970, Rondônia suffered a yearly caseload of 10,000 infections. By 1990, the disease was striking 250,000 times a year. Not only was the influx of new blood providing the parasite with more victims to infect, the potentially deadly disease had another unanticipated ally: the clearing of the forest.

Not all mosquitoes transmit the malaria parasite with equal efficiency. In the Amazon, the most deadly vector is the *Anopheles darlingi*. Shaped like a knife with wings, it has a hunger for human blood and a preference for open spaces. "Every species of mosquito has its special breeding sites," said Luiz Hildebrando Pereira da Silva, who heads the Tropical Medicine Research Center in Porto Velho. "Some breed in the vegetation. They put the eggs in the flowers or in the plant, so their reproduction depends on that. *Darlingi* likes large collections of water with sun."

When researchers in Peru compared forested areas with those that had been cleared, they found *Anopheles darlingi* mosquitoes bit nearly three hundred times as often outside the jungle. "We found a greater abundance of these dangerous mosquitoes and the mosquito larvae in deforested areas, even after controlling for human population," said Jonathan Patz, an environmental health scientist at the University of Wisconsin. "We sampled mosquitoes in abandoned agriculture fields. There were no people in the area at all, but we found these mosquitoes. At the other end of the spectrum, we also had a few village sites that were in the jungle, that is, dense human populations in a forested location. But there were much fewer of these mosquitoes, if any."

According to researchers at the center in Porto Velho, *Anopheles darlingi* makes up roughly 2 percent of the mosquito population inside the region's forests. Just outside, where settlers have their farms, the bloodthirsty insect accounts for 95 percent of the population. "He was living in the forest, but with difficulty," Pereira da Silva said. "Then man arrived. He cut the trees. He made large, beautiful collections of water. Ah, it was very good for *darlingi*. So he bred. He dominated."

The government of Brazil no longer actively promotes deforestation, but it remains ambivalent about slowing it down and continues to build and maintain roads, bring electricity to new communities, and supply settlers with infrastructure and services. Our drive took us through Monte Negro, once a center of the frontier economy, until the frontier moved on and the sawmills and the bulk of the population followed the forest west. The road onward was chewed and rutted. Twice, we were stopped while construction crews wrangled bulldozers, and covered tractor trailers wrestled with the slopes. The new center of the logging economy was a town named Buritis, a gray and dusty community of sixty thousand people, ringed with metal-sided sawmills and stacks of timber. On the roads beyond it, the trucks we met were smaller and uncovered, loaded with logs as wide as a man is tall. The landscape was still ranchland, but of a rawer kind. Clumps of forest clung to the hilltops. In the pastures, every few feet or so, the earth was split by burnt and blackened stumps—what remained of the Amazon forest.

Our road squeezed through clumps of glistening green. For a small stretch, it cut through untouched jungle. We had reached the most newly settled farms, the latest frontier in the conquest of the Amazon. Felled logs lay tangled in the fields. Stumps, some several feet high, marched across the earth, crossing several hundred yards of no-man's-land to the front lines where the forest still held control.

Jovelino Santino Dos Santos had arrived six months earlier at the start of the dry season. He had paid a little over eight thousand dollars to a squatter for rights to eighty-six acres of land. He was short and sinewy, with curly hair. His skin, bronzed by the sun, was tight on his face. When he smiled he showed a missing tooth. He hadn't shaved for several days. With a chain saw and fire, he had cleared about 8 percent of the land he had bought. Only the palm trees remained. Their trunks were too tough to cut. He estimated clearing the entire plot would take ten years.

Jovelino and his wife, Maria Aparaecida Dos Santos, had been weeding their cornfields when we drove up. They had walked up carrying a long length of firewood between them. It started raining, and we gathered inside their house. The walls were made from thin branches still covered in bark, with enough space between them for the light to shine through. The roof consisted of plastic sheeting over small beams of timber. On the far side of the curtain that divided the house in two, I could make out a matrimonial bed pushed up against a bunk bed for their two children. On our side, they had placed two small couches covered with cloth, a concrete hearth on wooden legs, a gas stove without its tank, and a bookshelf loaded with pots and pans. Baby chickens scratched at the dirt floor. The air smelled of wood smoke. Rain ran across the roof like a drum roll.

Jovelino was thirty-five years old. Before arriving, he had worked a small plot for a landowner in Buritis. "I don't plan to sell this land," he said. "I plan to stay here and work hard." Maria was thirty-three. She had named their new home Sol Nascente, Rising Sun.

A month earlier, Maria had come down with malaria. The next week, Jovelino caught a bad case. It laid him down for ten days. Then Maria again. Next was their thirteen-year-old daughter. The only one in the family who hadn't suffered from the parasite was their five-year-old son. "We can afford having malaria," said Maria. "But with these young ones, it's hard for them." The nearest nurse, a woman named Lurdes Sotinho, worked in a clinic in the village of Rio Branco, a trading center of unpainted woodboard houses a half-hour drive away. She served a population of roughly a thousand villagers and settlers and was seeing about fifteen cases of malaria a day. On average each person could expect to catch malaria five times in a year. "When people have a fever or vomit or any symptoms, they just come down to the clinic," she said. "Usually, when someone in the family catches malaria, the rest follow." I asked Jovelino if he had expected so much trouble with the parasite. His neighbors had all been sick. He didn't own a car, so each time someone in his family came down with malaria, he had had to ask for a ride. "I would have come anyway," he said. "If you want something special, to be a landowner, you must face this situation."

As the frontier moves on, malaria rates tend to drop in its wake. Breeding sites get filled. The Brazilian health care system strengthens its hold. Houses become sturdier, less friendly to mosquitoes. But even while the epidemic is abating, pressures are mounting for the settlers to relocate. In Rio Branco we had met several of Jovelino's neighbors, early colonists who had cleared a piece of land near Monte Negro or farther south, then sold it and moved again, starting over where the land was cheap and untouched. "The soils in Rondônia are not good for agriculture," said Matricardi. "But they can be very good for pasture—at least for the first ten or fifteen years. At the beginning it works very well. But then those pastures get old, so they start having problems. It becomes unsustainable, and so the small farmer moves. And then you start to see land concentration, people selling small plots, and the ranches become much bigger in area. Meanwhile, the small farmer is at a new frontier. He starts with some selective logging, selling those trees that are more commercial, and clearing land again, and getting malaria again."

IN THE SPRING OF 1993, MERRILL BAHE, A NINETEEN-YEAR-OLD cross-country track star from the Navajo reservation outside of Gallup, New Mexico, was driving into town with his family when he began gasping for air. He had had the flu, but suddenly he couldn't breathe. As he crumpled in the backseat, his panicked parents pulled into a convenience store to call for help. It was too late. Neither the paramedics nor the emergency room doctors could bring him back. His lungs had filled with fluid. He had suffocated.

It was the second case that Richard Malone, the medical investigator, had seen. Earlier that month, a thirty-year-old woman had died in the same emergency room, her lungs filled with a clear, yellow liquid. The cases had Malone worried and stumped, and he approached Bahe's parents looking for clues. "What they revealed gave him chills," Denise Grady wrote in *Discover* magazine. "Their son had died on the way to his fiancée's funeral, they explained. She had died five days earlier, with symptoms exactly like his. The couple had lived on the reservation with their infant son. But because the man's fiancée had died on the reservation, [Malone's office] had not been notified. 'That's when I realized we had a crisis,' Malone [said]."

By calling around, Malone and his colleagues soon identified seven cases. Six of the people had died. News of the deadly illness caused panic across the region. A week later, the brother of Bahe's fiancée and his girlfriend fell sick. She died, her lungs full of fluid. Most of the victims were less than thirty years old, strong and healthy, until fever and coughing suddenly kicked in. One had collapsed while dancing. Worst of all, nobody knew what caused the disease, or how it was transmitted. The press called it Navajo Flu. A few restaurants turned away Navajo families or served them with paper plates and plastic gloves. In return, some Navajos blamed white tourists for introducing the disease. Others whispered of poisons sprayed on peyote cactus or of weaponized germs from a nearby military base.

Scientists flown in from the Centers for Disease Control and Prevention in Atlanta worked in surgical masks and waterproof suits. A month after Bahe's collapse in the car, the death toll had reached fourteen, and another twelve people had been diagnosed with the disease. But the scientists had had a surprising breakthrough. Antibody proteins in the survivors reacted against the hantavirus family, a rodent-carried virus named for the Hantan River in South Korea. It was the first time a hantavirus had afflicted humans in North America. A year earlier, scientists at the Institute of Medicine had postulated that the Asian strains, which can lead to renal failure, could cross the Pacific. But the killer in New Mexico seemed to be a native virus. Instead of targeting the kidneys, it attacked the lungs. "We never dreamed of a scenario like this," Stephen Morse, a virologist at Rockefeller University, told Grady. "It would have seemed too close to fiction."

Robert Parmenter, an ecologist at the University of New Mexico, had been studying rodent populations in northern New Mexico since 1989. The previous fall, he had noticed an explosion in the population of deer mice. In areas where Parmenter would normally catch three mice, his traps were snaring thirty. "When this was identified as a rodent-borne disease, it was pretty clear that there was a high population density of these animals at the time the outbreak was occurring," he said. More mice meant more animals living closer together, a perfect scenario for the virus to spread. "Once the population builds up, the hantavirus can move through it rather quickly," said Parmenter. The virus had almost certainly been present in New Mexico as long as the deer mice, but it had never reached levels large enough to have caused a detectable outbreak.

"The hantavirus doesn't actually hurt the mouse," Parmenter said. "It's transmitted to people through aerosolized urine and fecal material. Basically, you have a lot of mice living in a garage or a woodshed or underneath the floor of a house. Humans in the spring usually get in there and start cleaning up these places. They're using brooms or vacuum cleaners, and they stir it up into the air. They're usually in a closed dead air space inside a building or underneath a trailer, and they're inhaling this aerosolized rodent urine that has virus in it. That fit perfectly well with standard epidemiological theory. So the question became, why are the rodents abundant now?"

The answer lay in the climate. The year of the outbreak and the previous one had been El Niño years, when the Pacific Ocean off the coast of South America warms up, disrupting weather patterns and showering the southwestern United States with rain. "That had promoted an increase in food abundance for rodents—vegetation and insects and berries and nuts and acorns," said Parmenter. "The population grew very quickly, and by the spring of 1993, it was just unbelievably high." When El Niño returned five years later, Parmenter had a test for his theory. Between 1993 and 1997, just one in ten of the mice he captured carried the virus. After another unusually wet season in 1998, the population soared, and half the animals he tested were infected. "Typically in New Mexico we might have four human cases per year, but following the wet years, when conditions are great for the mouse populations, we can double or triple that number," he said. "In really dry years, when the mouse populations are low, we might have only one or two cases, sometimes zero." Parmenter had linked the hantavirus's emergence to oscillations in the global climate.

If alterations in the local landscape or disturbances in the weather can lead to unexpected outbreaks, we can expect global warming to do the same thing. Most diseases are found in the tropics, the hottest part of the earth. As temperatures rise across the globe, their range will spread. "As the climate starts directional changes, that's going to drive the water cycle," said Parmenter. "Some places will get wetter. Some will get drier. Sea level rise will influence where marshes are. Anytime you have a vector-borne disease, that vector is going to be susceptible to climate change. What you'll see is that species ranges are moving north and upslope."

"Birds right now are changing their distribution," said Parmenter. "Even in my house, here in New Mexico, for the first eight years I lived here, we never saw a white-winged dove. Now I see them all the time. White-winged doves are normally a southern New Mexico species, but they're coming up. You see that sort of thing all the time. Has it been documented in rodents? No, not yet. But I'm sure it will be. Anytime you change a habitat and environmental conditions in a way that favors the vector of a disease, sooner or later you're going to start seeing an increase of that disease."

After solving the mystery of the hantavirus, Parmenter

turned his attention to the bubonic plague. Again, it was weather that drove the disease. In the hot deserts of New Mexico, rainfall made all the difference: wetter winters and springs led to jumps in the number of cases as rats and fleas proliferated. Meanwhile in Kazakhstan, where the climate is much colder, scientists were discovering that warming weather is the determining factor. Research led by Nils Christian Stenseth of the University of Oslo found that a rise in springtime temperatures of 1.8 degrees Fahrenheit pushed up the prevalence of plague by nearly 60 percent. When Stenseth compared his findings with historical temperatures, he found that history's major outbreaks had occurred in years when Kazakhstan was warmer and wetter. "Our analyses are in agreement with the hypothesis that the Medieval Black Death and the mid-nineteenth-century plague pandemic might have been triggered by favorable climatic conditions in Central Asia," he concluded.

In Nigeria in 2002, I caught malaria twice. I cured it early both times, before it had time to gather strength. Nonetheless, it knocked me down. Walking to the clinic for a blood sample during the second attack, I vomited on the road. That night, on a friend's couch, my bones ran cold, my skin shivered, my shirt changed color with sweat. I lost two days before I was ready to work again.

As temperatures increase, regions that once were free of disease-carrying mosquitoes will become suitable for the insects. "We're seeing changes in the range of mosquitoes and the diseases they carry," said Paul Epstein, associate director of the Center for Health and the Global Environment at the Harvard Medical School. "That's going to be an issue increasingly in terms of latitude and on the margins, both in terms of extensions of the range and in terms of seasonality."

Mosquito larvae mature more rapidly when the water in which they grow is warm. Female mosquitoes digest blood faster and bite more frequently as the mercury rises. Malaria transmission begins when a mosquito feeds on an infected person. Once ingested, the parasites split into males and females, reproduce in the mosquito's gut, and release snake-shaped sporozoites that migrate to its salivary gland, ready to be injected when the insect bites a new human. A malaria mosquito will only live a few weeks. The parasite's survival depends on it reaching maturity while its host is still alive to bite. The strain of malaria I caught is called *Plasmodium falciparum*. In temperatures of sixty-eight degrees Fahrenheit, the parasite takes twenty-six days to complete its reproductive cycle. At seventy-seven degrees Fahrenheit, it's ready to reinfect after just thirteen days.

Climate change is also accelerating the spread of dengue, a primarily urban, tropical disease that causes fevers and aching. The disease has reemerged in Brazil and climbed up the American coast. In Mexico, the number of cases has risen 600 percent since 2001. Outbreaks have reached the formerly dengue-free state of Chihuahua, on the border with Texas. The Intergovernmental Panel on Climate Change estimates that by 2080 an extra two billion people will live in areas that are hospitable for the virus.

Scientists studying the Ebola virus in Gabon say that outbreaks erupt after wet weather breaks a long, dry spell. In North America, the spread of the West Nile virus was accelerated when warm temperatures and drought in 1999 favored the carrier mosquito and the disease within it. That same year, a new virus emerged in Malaysia. Deforestation and drought-driven forest fires drove jungle bats to feed in orchards near where farmers kept their pigs. The animals fed on the bat droppings, picked up the Nipah virus, and passed it on to their keepers. One hundred and one people died, and nearly a million pigs were culled, but the virus escaped to Indonesia, Australia, the Philippines, and Bangladesh.

In August 2007, an epidemic swept through Castiglione di Cervia, a small village in northern Italy. More than one hundred of the town's two thousand residents came down with high fever, rashes, and crushing pain in their bones and joints. An unusually mild winter had allowed Asian tiger mosquitoes to start breeding early, and their population had soared. When an Italian tourist returned from India with chikungunya, a relative of dengue, the insects provided the perfect vector. It was the first time the disease had broken out on the continent. "By the time we got back the name and surname of the virus, our outbreak was over," Rafaella Angelini, director of the regional public health department in Ravenna, told the *New York Times*. "When they told us it was chikungunya, it was not a problem for Ravenna anymore. But I thought: This is a big problem for Europe." According to officials at the World Health Organization, the epidemic was the first European outbreak of a tropical disease caused by climate change.

"This is all part of a pattern of emergence of new diseases and of the resurgence and redistribution of old diseases," said Epstein. "It's occurring globally for many reasons: deforestation, changing habitats, chemical use that affects predators, our own use of antibiotics, and then climate." Disruptions in normal weather patterns favor opportunistic pests and parasites. Climate change has the potential to unsettle ecosystems and bring humans, animals, and pathogens together in new and unexpected ways. Since 1976, the World Health Organization has identified thirty-nine previously unknown diseases, including Ebola, the New Mexican hantavirus, and Lyme disease—an eruption of pathogens on par with those that occurred during the invention of agriculture and the Industrial Revolution. "As the climate becomes more unstable, it will have an increasing impact," said Epstein. "We're going to see things shift."

WILLIAM RUDDIMAN WAS NEARING RETIREMENT WHEN HE CAME across a graph of historical methane levels. A paleoclimatologist at the University of Virginia, Ruddiman had spent his early career examining ocean sediments for hints of ancient temperatures and focused his later research on the major drivers of the world's climate. But the data, gleaned from air bubbles trapped in the Antarctic ice sheet, surprised him. "I had a very clear expectation of what those concentrations should have been doing," he said. Monsoons in the tropics have been weakening for the past ten thousand years, shrinking the swamps and wetlands that create most of the world's natural methane. The trend in the ice cores should have been unrelentingly in decline. "It does go down from ten thousand to five thousand years ago," Ruddiman said. "But then it reverses direction and goes up. It goes so far up that by the time we get to the industrial era, it's as if the monsoon is cranking away at full speed. And yet the tropics are drying out. The methane trend went the wrong way. It shouldn't have done that. Why did it do that?"

Ruddiman's hypothesis, published after his retirement, generated more controversy than anything he had done during his career. "If the natural sources weren't putting this extra methane in the atmosphere, then it must have been something else," he said. "Humans were the obvious possibility." The bend in the methane graph, he concluded, occurred when farmers in China first began to grow rice in large quantities. "People started to irrigate," he said. "They were better fed. The population started to boom. Larger numbers of people kept larger numbers of livestock, and that's a source of methane. The people themselves are a source of methane. They probably burned off their fields every year, and that's another source of methane." Human civilization, Ruddiman was arguing, had been altering the global climate since the beginning of large-scale agriculture.

When he looked at carbon dioxide, Ruddiman found a similar trend. "It should have been falling the last ten thousand years, up until the industrial era," he said. "And it did fall for a couple of thousand years, and then it started to rise. It's a parallel problem. Why did it go up when at similar times in the past it's always gone down?"

"Well, there's really good data in Europe showing the beginnings of what would become major deforestation starting just about eight thousand or seventy-five hundred years ago," he said. "And by the time you get to maybe a thousand years ago, most of Europe is completely deforested. There's probably more forest now in most of Europe than there was then." Intriguingly, carbon levels didn't rise steadily. Around the time of the Roman era, they wobbled, dipped, then swung back up around A.D. 1000. Shortly after A.D. 1500, in a period coinciding with the Little Ice Age, when the northern hemisphere plunged through a series of bitterly cold winters, they dove again. If the rise in carbon dioxide was due to population growth and the cutting of trees, could the wobbles and drops Ruddiman was seeing have been caused by depopulation and reforestation?

The early dips matched up with plagues during the Roman and medieval eras, but the dating from the ice core data was hazy. Ruddiman concentrated on the biggest decline, the one that began after A.D. 1500 and lasted more than two hundred years. "That carbon dioxide drop correlates with the biggest pandemic of all of preindustrial history, the arrival of Europeans into the Americas," he said. Before the arrival of Christopher Columbus, the New World was home to between fifty and sixty million people. Two hundred years later, around five million native Americans remained.

"Entire villages that once lined the valleys of the lower Mississippi River system were abandoned, along with endless cornfields in between," Ruddiman wrote in *Plows, Plagues, and Petroleum: How Humans Took Control of Climate.* "After the forests again took over, the only obvious evidence left of the former existence of these agricultural people was massive earthen mounds used for ceremonial purposes, and most of these mounds were plowed by settlers and flattened to create towns and cities. In the Amazon Basin and other rain forest regions, lush tropical vegetation swallowed up most evidence of former habitation. Many decades later, so little evidence remained of the former occupation of North America that scientists and historians in the 1800s and early 1900s assumed that populations had been relatively small."

Smallpox, typhus, cholera, measles, and a host of other diseases had killed roughly one-tenth of the earth's population. The trees that reclaimed their farms and villages sucked the carbon out of the air. "How much did the American pandemic contribute to the Little Ice Age?" said Ruddiman. "Conceivably all, but at least half, of the carbon dioxide drop that occurred is due to the pandemic."

"There's a tremendous amount of carbon stored in the biomass of the forest, also in the soil," said Philip Fearnside, a professor at the National Institute for Research in the Amazon in Manaus. Roughly one hundred billion tons of carbon are tied up in the leaves, branches, vines, trunks, and roots of the Amazon, more than a decade's worth of the world's fossil fuel emissions. Every time the forest burns, some of that carbon is freed into the atmosphere. In Brazil, deforestation produces more emissions than all the cars and industries combined. The country releases about eighty million tons of carbon into the atmosphere by burning fossil fuels. In 2006, the amount produced through burning trees, rotting logs, and the decomposition of humus in the earth amounted to three times that amount.

The Amazon is contributing to what may be its own extinction. Climate change models forecasting the future of the forest differ on what will happen in equatorial South America. But the model that best replicates the historic drought cycles caused by the globe's warming oceans predicts massive drying: decreasing rainfall that wipes out the Amazon within seventy years. "The fires aren't even included in these models," said Fearnside. "It's being wiped out just from the trees dying of thirst."

To get an idea of how the jungle responds to drought, Daniel Nepstad, a scientist from the Woods Hole Research Center, in Massachusetts, dug a deep trench around a hectare of Amazonian forest and covered the forest floor with more than five thousand plastic panels. Gutters channeled 30 percent of the water into a nearby valley. The forest held up under the artificial dryness for three years and then started to die. "The trees that were most susceptible to drought-induced death were the big trees," said Nepstad. "That forest is damaged for many years afterwards. You're taking the dominant organisms of the forest and killing them. They go crashing to the forest floor as they die, opening up new gaps in the forest. Basically, you're taking a very fire-resistant ecosystem and making it into a fire-vulnerable ecosystem."

Nepstad combined his findings with economic models for the future of Brazil's logging and agricultural industries. In comparison, the dramatic drying predicted by the climate models looks optimistic. By 2030, Nepstad predicted, farmers will have deforested nearly 31 percent of the remaining Amazon. Logging will thin another 12 percent. Drought will degrade another 12 percent, more if climate change cuts rainfall as expected. More than half the forest will be gone or damaged. Twenty billion

tons of carbon will have been released into the atmosphere.

What would an Amazonian die-off mean for farmers like
Jovelino? If deforestation by chain saw and burning promotes malaria, will the loss of trees through drought and
forest fires do the same? "If the forest shrinks, what's going
to happen to malaria?" said Ulisses Confalonieri, a professor of public health at the Oswaldo Cruz Foundation in
Brazil. "Is the Amazon going to become a savannah type of
vegetation? We don't know if the mosquito is going to per-
sist in those areas or not." In Africa, where the disease is
seasonal, transmission of the parasite depends on the
interaction between the weather and the long-term climate. In wetter zones, caseloads rise during dry spells—
when mosquitoes can reproduce in stagnant rivers—and
drop when showers flush away the breeding sites. In areas
that are normally dry, the pattern is reversed. Rains create
puddles in which the insects can breed. Parched periods
mean few mosquitoes and lower malaria rates. In the
Amazon, infection rates tend to drop slightly during years
of drought. But research also suggests that dry weather
allows mosquitoes to breed in the slow-moving water of
the rivers, near which most of the region's population lives.

Such uncertainties will pose climate change's biggest
challenge to public health. Diseases are easier to fight
when you know when and where they'll hit. Outbreaks
find opportunity in times of uncertainty. As with West Nile
virus, increased numbers of mosquitoes and other vectors
will help viruses and parasites evade the sentinels. Formerly

malarial regions like Italy, France, Florida, and parts of Brazil, where the parasite is kept in check through spraying breeding sites and the quick treatment of the sick, will come under increased pressure. "The limiting factors are the climate situation and the vigilance of the public health officials in the area," said Luiz Hildebrando Pereira da Silva of the Tropical Medicine Research Center in Porto Velho. "If you have conditions for improving the density of mosquitoes, you will increase the number of epidemics. This will increase responsibility and the work for the sanitary control."

Droughts, natural disasters, and climate-caused conflicts will disrupt health systems and put people and epidemics on the move. "One important issue in terms of climate scenarios is for the Brazilian northeast," said Confalonieri. "It's a semiarid area. If the temperature increases and the rainfall decreases, we're going to have a mass migration of people out of the region." During periods of droughts caused by El Niño in the 1980s, ruined farmers in the region fled to cities in search of work and triggered unexpected outbreaks of kala-azar, a potentially deadly disease that attacks the spleen. They had come from endemic areas and caused epidemics where the disease was unheard of. Other farmers escaped to the nearby Amazon, caught malaria, and, when the drought ended, brought the parasite home to their neighbors.

In 2005, during the Brazilian drought, hundreds of communities in the western Amazon were cut off from health care when dried up lakes and rivers meant they could no longer use their boats. Melting permafrost in places like

Russia will have a similar effect. The World Health Organization estimates that at the turn of this century global warming was responsible for three out of every one thousand deaths. In 2000, it killed 150,000 people, through heat waves and disease. In 2003, high temperatures in Europe killed as many as 45,000 people in two weeks. Global warming is responsible for 2 percent of malaria cases and one in forty cases of diarrhea, a major killer of children in the developing world, according to the World Health Organization. Rising pollens and molds team up with desertification and forest fires to boost respiratory infections, allergic diseases, and asthma rates. Droughts like the one in Darfur lead to malnutrition and death. If current trends continue, the death toll attributable to climate change is expected to double in the next thirty years.

One evening in Porto Velho, I joined four workers from the state's malaria control program for an evening drive into the nearby countryside. We left the asphalt just out of town, gunned up a steep dirt road, and stopped at a sheep farm. A small wooden shack was furnished with bales of hay and a small color television. Two sides of the porch were turned into sheep pens. On the third, clothes hung from a line. Geese waddled in the mud. The sun had begun to drop, and my companions fanned out and got to work. Each had a short stool, a flashlight, cups lidded with mosquito netting, and a long rubber tube. I followed their leader, a short, dark-haired man named Ernaldo Cunha Santos. He rolled up his pants, set his stool next to the shack, pulled his black cotton socks to his knees, and waited for the mosquitoes to bite.

Each time one landed he would spot it with his flashlight. With the rubber tube in his mouth, he would suck it up and blow it into a cup. Twenty minutes later, he showed me his catch. The cup was swarming with mosquitoes. Lean and hungry, they zipped like darts from wall to wall. Santos's tube swept across his leg, vacuuming up five at a time. He paused and gave me a look of mock forbearance. Then alerted by a sudden itching in his ankle, he turned his attention to his socks.

Even where I stood, mosquitoes were biting my wrists and knuckles. After about forty-five minutes, I asked Santos how many he had caught. He gave me a thumbs-down. It had rained in the afternoon, and the mosquitoes weren't biting as much as usual. He had caught only 120. We piled back into the truck and drove back to Porto Velho. Together, the four men had been bitten by 390 potentially malarial mosquitoes in under an hour. The collection was a daily ritual.

The next morning Santos and his colleagues would test the insects with pesticides so the state could adapt to any signs of resistance. The economic cost of malaria is high. The disease is both a cause and an effect of poverty. In Africa, where it is most rampant, the World Bank estimates epidemics cost the continent $12 billion a year and slow economic growth by as much as 1.3 percent. In the countries that are hardest hit, even simple prevention measures like five-dollar bed nets—let alone antimalarial drugs or robust mosquito control—are out of reach for the very poor. "Where malaria prospers most, human societies have prospered least," wrote Jeffrey Sachs, director of the Earth Institute at Columbia University, and Pia Malaney, an economist at Harvard, in *Nature*. "The extent of the correlation suggests that malaria and poverty are intimately related."

The health care systems in richer countries are likely to be able to dampen the effects of climate change on the spread of disease, but countries in the third world will suffer the full consequences. "Restrictions on the movement of goods could also be a source of economic and political turmoil," wrote John Podesta and Peter Ogden of the Center for American Progress, a progressive think tank, in the *Washington Quarterly*. "Pandemic-affected countries could lose significant revenue from a decline in exports due to limits or bans placed on products that originate or transit through them. The restrictions placed on India during a plague outbreak that lasted for seven weeks in 1994 cost it approximately two billion dollars in trade revenue. Countries that depend on tourism could be economically devastated by even relatively small outbreaks."

Vector-borne diseases will spread through the highlands of poor countries like Kenya, Uganda, and Zimbabwe and chip at the progress made by emerging economies. Subhrendu Pattanayak, an economist at RTI International, a nonprofit research corporation in North Carolina, argues that countries in the Amazon need to prevent deforestation if only to blunt the cost to the economy as malaria rises due to climate change. "Sure we have an Africa story," he said. "But there are countries like Brazil and India and Indonesia and Malaysia which are not so poor, where the consequences could be pretty serious." In the battle against disease and climate change, Pattanayak said, it's the middle-income countries that have the most to lose. "Some of these are about to take off and become major global powers," he said. "But they're still susceptible to these outbreaks. That's what I would worry about."

5

"BEAUTIFUL COUNTRY"

THE WEST COAST, HOTTER SUMMERS, AND THE GRAPE HARVEST

Cain Vineyard and Winery sits in the mountains between the Napa and Sonoma valleys, up a series of steep, narrow, wooded roads from the tourist tracks below. There's no sign at any of the turnoffs and only a small one at the property's edge. Visitors are welcome, but not sought after. Tours are by appointment only, on Friday and Saturday mornings when most of the staff presumably don't have anything more urgent to take care of. The estate's tasting room is large and opulently decorated. On the day I visited, a small table was set for three, with wineglasses, disposable cups, and white paper place mats. Above an unlit fireplace hung a painting of cowboys riding their herd across a desert plain.

I was joined by Chris Howell, the estate's winemaker, and Ashley Anderson, its vineyard manager. Both were dressed for the fields. Howell wore a loose-weave, short-sleeved shirt, faded blue by the sun. His khaki pants stopped a bit short, above brown leather shoes. His voice was a touch nasal. Anderson was dressed in a long-sleeve yellow shirt and blue jeans. The orange of her baseball cap matched the color of her work boots. She had long brown hair and a worker's tan.

We began with the previous year's Cain Cuvée. Howell uncorked the bottle and poured three glasses. "What we've got here is a red wine," he said. "The color comes from the skins." He held it up at an angle, examining its hue against the place mat. "The juice this comes from obviously has sugar and acid," he said. "But the key is the perfume and the aroma. What's the difference between a great peach and a so-so peach? It's all about the perfume." He put his nose into the glass, took a deep smell, lifted it to his lips. I did the same. A bitter taste gave way to the tang of pine, then to a rich herb like dried oregano. We spat into the paper cups.

"What do you think, Ashley?" said Howell. "Ashley's a better taster than I am."

"I get some spicy notes," said Anderson. "Not really green spicy but sort of herbal, cooking spices."

"Yes, tarragon," said Howell.

"And kind of green forest," said Anderson.

"The goal is, how does it go with food," said Howell. "How does it linger? What's the overall impact, and what's the balance?"

Howell poured a second type of wine, a blend called Cain Concept. "The difference between these two wines is primarily where these grapes grew, which vineyards," he said. He held up his glass. "Darker in color," he said. "Darker doesn't necessarily mean better, but it is darker in color. On the nose, I think it's less herbal and more sweet fruit, a little less red pie fruit and more black cherry fruit."

"It's more voluptuous," said Anderson.

"And the entry is a bit sweeter, rounder, fuller," said Howell. "The texture, the tannins, are smoother."

"Yeah, the mouth feels a lot different," said Anderson. "It's rounder. It's silkier."

"The only point of all this is to say, 'Yeah, these wines are all different,'" said Howell. "There may be no other thing grown where we're so sensitive to differences in tastes. We may, if we have our own garden, say, 'This is not such a good year for tomatoes.' But most people don't think about that. They just go to the store and buy a tomato."

I had come to Cain to speak with Howell about the effects of climate change on the wine industry, and the impromptu tasting was his illustration of the impact small differences could make. Howell subscribes to a classical European view of wine growing that holds that the determining factors in the production of a quality vintage are the varieties of grapes used (cabernet, pinot noir, merlot, and so on), the soil from which they draw their nourishment, and the climate in which they grow. By the time a fine wine is finally uncorked, the way it tastes has been formed by a nearly infinite series of choices: How were the grapes crushed? Which strain of yeast was used? How long did the pulp, skin, and seeds settle? How and when was it filtered? At what temperature did it ferment? Did the winemaker add acid or sugar? What forests produced the oak in which it matured? How was the bottle stored and for how long?

But in the end, the thing that really matters is the quality of the grapes. A clumsy winemaker can turn fine fruit into bad wine, but a top vintage can't be made from anything but the most exquisite pickings. "Every winemaker

really knows that the bottom line is the grapes themselves," said Howell.

Yet grapes, especially those used to produce fine wines,
are particularly susceptible to changes in the climate in
which they grow. The top vintages are made from grapes
that mature at just the right moment in the season and at
just the right speed. If it's too hot, the sugars in the fruit
rise too fast, leaving the flavor compounds—the elements
that give the wine its complexity—with no time to develop;
the winemaker is forced to pick early before the fruit over-
ripens. Not enough heat during the growing season and
the grapes reach the harvest bitter, with their seeds still
green; the grower must choose between cutting the ripening short or risking autumn rains and moldy, bloated
grapes. Unlike with other fruits, which need to be mar-
keted, distributed, and displayed in ways that catch the eye
of the buyer, the only thing that matters when choosing
when to harvest grapes is what you'll get when you open
the bottle. "We get to pick our grapes not based on what they
look like, or how far they can be shipped, or what they
might taste like in three weeks, but on what they taste like
at that moment," said Howell. "What's wine about? It's
only about flavor."

Grapes are the only crop for which consumers want to
know the year in which it was harvested. The difference
between a highly rated vintage and a lesser one doesn't
usually depend on changes in the variety of grape that was
pressed, the soil from which it sprung, or the techniques in
the production. In most cases, the only thing that varies
from one year to the next is the weather in which the
grapes grew.

. . .

As the climate changes, so will the way wine is grown. Given a patch of land with the vines already planted, the craft boils down to making sure each grape gets just the right amount of heat. On the wall in the tasting room, Howell had hung aerial pictures of the Cain estate. Shot in infrared, his vineyards showed up as crimson quilts. Brighter hues reflected heightened levels of chlorophyll, broader leaf surfaces, and more vigorous grapes. A streak of deep red running perpendicular to the rows of vines marked an underwater stream. "That can help us with taking decisions," said Anderson. "We know that in this area the grapes are most likely to ripen up differently than in this area which is right next to it. We'll harvest these ten acres in at least eight different picks."

Earlier in the day, Howell had taken me on a tour of the vineyards, accompanying the drive with a continuous commentary on microclimates and *terroirs:* how the morning sun hit one site, how the soil changed in character. Howell is obsessive enough that he asks the same worker to prune the same vine each year to eliminate any influence from a difference in styles. As with most perfectionists, his was primarily a tour of his defects. He had planted rows of grapes on terraced hills, two to every step, and had noted different ripening on the uphill and downhill sides. In one spot, heat shimmered off the vines. The plants had burned, he said, and lost their leaves.

On the slopes, the job boils down to ensuring the grapes ripen at the same rate and finish at just the right time. Almost anything can make a difference: whether a hill faces north or south, how many clusters are left on each vine, the spacing between the trellises, how and when the vines are cut back, how much breeze blows through the leaves, the quantity and timing of the watering. "It's all about the cluster," said Howell. "Where is it in the vine? What's the sunlight? What's the heat? What's the airflow? You can talk about the temperature down in the valley all day long, but it's the temperature of that cluster that's going to affect what the wine tastes like."

We followed an unpaved road up to a ridge at the property's edge. We were at a break in the watershed between the two valleys. Howell's vineyards lay on the Napa side of the crest. Sunlight glared off fields of dry golden grass. From the other side, where Sonoma began its climb down the mountain, a cold wind blew. Further up the ridge, a single tree bent with the gusts.

Howell opened a chain-link gate, and we stepped down into the vines, careful not to slip on the gravel slope. Rusted steel posts supported a thick wire trellis, to which the vines clung as they grew. Along the bottom, at about shin height, black irrigation pipes slithered through the rows. The trunks were brown and knotted, about as thick as a child's arm. It was a sunny summer day, not long before the harvest. The vineyard workers had been busy dropping fruit, culling the clusters that had been slow to ripen. Bunches of discarded cabernet berries shriveled at the base of the vine, black against the golden grass.

One way to control the temperature experienced by a given bunch of grapes is to manage the layout of the vine. The clusters can be placed closer or farther from the ground, which alters the amount of reflected and radiated heat that reaches them. Leaves can be trained or pruned to regulate sunshine and air currents. Howell squatted by the grapes. The bunches hung at about the height of my knee. The leaves were relatively sparse and pulled up against the plane formed by the row of vines. "This is the coldest place in our vineyard," Howell said. "My goal fifteen years ago was to get more heat right here. I thought it was so darn cold I wanted to train my vines closer to the ground. And it was absolutely necessary to have a very neat vertical system, where these clusters have no shade at all." The patch was one of the first Howell laid out when he took charge in the 1980s, fresh from studying and working in France. "What I discovered was that there's too much heat, that this system—which evolved in Europe, in a much rainier climate and a much cloudier climate—wasn't appropriate here," he said. "I'd get too much sunlight on this cluster, too much heat from this soil, and the fruit would literally shrivel before it was ripe."

"Now, we let it grow ragged," he said. "We let shoots stick out, to create a little shade. And as we evolve our planting, we've put the head height progressively higher and higher."

Earlier, Howell had taken me through a metal gate behind his house that opened on rows of ripening vines. The rows he was showing me were the oldest on the property, trained in a traditional style called California sprawl. The clusters lay about halfway up the vine. The leaves were allowed to shoot in different directions, providing shelter from the sun's hottest rays. "What you've got here is the clusters apparently sitting in the shade," he said. "But if you look closely there's little bits of sunlight. There's very little foliage really burying the cluster. You can see every cluster, and that's the big deal. At some point during the day every part of this cluster is getting a bit of sun."

He pulled at the tangle of leaves that formed the canopy. The grapes underneath were the size of marbles. Dusty purple, they grew in small bunches. "Taste one of these," he said. The berry was small and soft between my fingers, with none of the bulk, shine, or tautness of a table grape. I dropped it into my mouth and broke it open with my tongue. The juice was sharp, with a spike of sour apples. The seeds were thick and coarse. "Chew the skin," said Howell. "It's a little tannic, a little dry, mouth cleansing, palate cleansing."

He waited while I finished, then lowered the canopy back over the fruit. "This is a system of growing grapes that I used to think was silly," he said. "And now I think it really works great," said Howell. "It's definitely adapted to warmer temperatures and to a very sunny, cloudless climate."

In the tasting room, I asked Howell what a rise in temperatures would mean for the wines he was making. The two vintages we had tasted had been blended from grapes grown within a few miles of each other—they had experienced small changes in variety, soil, and climate—but the Concept retails for twice the price of the Cuvée. "The essence is that if anything changes, it changes the status quo," said Howell. "If climate change means it gets warmer, the wine is going to taste different. Period."

"You can use the personal experience you have with fruit of any sort, a berry or a peach or an apple or a plum," he said. "There's a moment of unripeness and then ripeness.

And then there's the stage where you say, 'If I'd gotten this last week it would've been really, really good, but it's now soft and mushy and not quite the same flavor.'"

He picked up his glass. "This wine, for many people, it could get riper," he said. "But at some point it would get overripe. Those aromas are changing; they're moving from bell peppers to tobacco to cedar to blackberries to black cherries to plums to dried fruit. And then like that peach, it would lose all of its perfume. It would lose its aromas, and in fact the color would fall away. And then everybody would agree it's getting worse."

"We have to adapt," he said. "In fact, it might be that we have to say we're not growing grapes here anymore. That's a possibility. That's not what my marketing group would like us to say. Will it happen? It could. Will we still have good wines? I think in the near term, yes. In the long term, I don't know."

When I first arrived, Howell had walked me to the tiled terrace behind his house. We stood by a small swimming pool and a hot tub and took in the view. On our left, the mountain dropped steeply across wooded slopes. A summer haze hung in the air. Through a cleft in a bluish hill, we could see the reservoir pool for the town of St. Helena. To our right, the vineyards awaited the pickers. Howell's grape-laden vines etched contours into the rolling landscape.

"Beautiful country," I had said.

"Beautiful country," he had answered.

"It's like in those serial killer movies," he had said. "'What a beautiful face: too bad we'll have to ruin it.' This is a beautiful place. I just hope we can even grow flowers here in twenty years."

The relationship between weather and wine growing is reliable enough that climate scientists have begun to gather historical harvest records to get an idea of the climate before the invention of the thermometer. The date on which the picking begins, it turns out, is a reliable measure of how hot the summer was. Picking dates in France have been set by decree since the Middle Ages, decided by the local authorities about a month before the harvest is to begin.

Pascal Yiou, a climatologist with the Laboratoire des Sciences du Climat et de l'Environnement in Gif-sur-Yvette, France, began in Burgundy, choosing the French wine-growing region for the continuity of its records and for the fact that almost all the grapes grown there are of the pinot noir variety and thus would ripen at the same rate. By combing through municipal archives, Yiou and his historian colleagues assembled uninterrupted records of harvest dates going back to 1370. The resulting temperature data matched up nicely with records from the modern era and tracked with estimates gleaned from tree rings in central France. "We can see the entrance into the Little Ice Age, and we see it warm at the end of the nineteenth century and the continuous warming in the twentieth century," Yiou said. A graph of their data shows several historical summers hotter than those of the 1990s, but the 2003 European heat wave, one of the hottest summers on record, spikes high above the rest. "The harvest date in

2003 was the earliest on the record by far, by a few standard deviations," said Yiou. "People harvested in mid-August. This has never been seen since the records began."

While scientists were noting the early harvest date, the French wine industry was aghast at what the heat had done to their product. "That year, 2003, was the wake-up call," said Jancis Robinson, a wine writer and columnist for the *Financial Times*. The early harvest—in the dead heat of summer—had growers scrambling to get the grapes off the vine. Many slept in their wine cellars; it was too hot in their houses. Some rented refrigerated trucks to cool down the grapes as they came out of the vineyards. "The trouble was that the grapes dried on the vine," said Robinson. "They became raisins. Their sugar levels were not accompanied by the nice steady development of phenolics and other interesting things. It was almost as if they were arrested halfway through the ripening and the gaining of complexity."

We spoke in Robinson's garden, behind her North London home. Her blond hair was cut short. The brown rims of her eyeglasses matched the color of her tights and her leather shoes. The red on her lips was the same as that of her dress. "If you taste an '03 French wine they're still tasting reasonably nice now, because they're very rich," she said. "But at the end you can taste that there wasn't enough juice in the grape. The taste is slightly raisiny."

"I don't think they've got enough ingredients to make them interesting after ten years," she said. "Apart from the very best ones, that dryness will get more and more prevalent." In many cases, the best vintages that year came from lesser-known estates, ones that historically had struggled to ripen their grapes. In a normal year, their vineyards were too shady, too windy, too cold. In 2003, they had enjoyed just the right climate. "That summer, the bad thing was to be trapping all that sunshine," said Robinson. "If we get more of these hot vintages, we'll start actively looking for these lesser vineyards."

In addition to contributing to newspapers and writing for her Web site www.jancisrobinson.com, Robinson edits the *World Atlas of Wine*. She had spent the week before our meeting putting the final touches on a new edition. Many of the updates, she told me, reflected the effects of global warming. "There are areas at the polewards extremes of the world's wine map that used to have to struggle to ripen their grapes," she said. "There are whole countries that had just a madman with a couple of rows of vines. Belgium, Poland, even Denmark, they now have domestic wine industries. The wines are not going to knock anyone's socks off, but they're good enough to sell in the local restaurants. Germany used to not be able really to make red wine. The grapes wouldn't get enough pigments in the skins. They would have to be picked too early, before flavor could be built up. Now the second-most-planted grape variety in Germany is pinot noir. The Burgundians always pride themselves on having quite a continental climate. There used to be a time when they all went on holiday throughout August because they knew they were never going to pick before September. Now all leave is canceled because the grapes are ripening so fast."

"There was a general assumption then that 2003 was a one-off," she said. "Everybody thought that this would never happen again. But 2005 was pretty hot in the summer too. There's been a gradual realization that it wasn't a flash in the pan. They will get other years like 2003."

That same year saw unusually hot weather in the American West. Drought and strong winds fueled devastating wildfires in Southern California. In the Howell vineyards above the Napa Valley, temperatures never rose very high, but the growing season started early. Farther north, Oregon's Willamette Valley, a generally cool region known for its pinot noir, was having the hottest summer in decades. I met Harry Peterson-Nedry at an outdoor restaurant not far from his vineyards, a pinot noir estate called Chehalem. He wore a blue and yellow striped short-sleeved shirt, beige pants, and loafers. His hair was white. A former chemist who designed magnesium batteries to withstand the Vietnam heat, Peterson-Nedry had begun dabbling in wine twenty-seven years earlier. "It was one of the few things that bridged the right brain and the left brain," he said. "Wine has the hedonistic sensibility side of things, and it's also got the rational, scientific side."

As we ate lunch, Peterson-Nedry showed me a graph on which he had plotted the heat accumulation for the past ten seasons. Grapes don't ripen until the temperature reaches about fifty degrees, so he had plotted the time that the grapes had spent above that threshold, a rough estimate of the amount of useful heat they had accumulated, and thus how fast they had matured. Each year was shaped like a mountain in profile: a slight slope in spring, a steep rise in summer, and a slow leveling off in the fall. As in France, 2003 marked the high-water point. And 2006 wasn't far behind. The graphs also included the average for the heat accumulated each year between 1961 and 1990.

"Ten out of the last ten years have been at that average or above," Peterson-Nedry said. "The likelihood of this having happened by chance, rather than actual climatic change, is about two out of a thousand."

In addition to using the graphs to monitor the warming in the valley, Peterson-Nedry used them to plan his harvests and his plantings. "In 2006, we knew it was a year like '03," he said. "We had a decent crop load that was set during June. Normally we go in and drop anywhere from a third to a half on the ground, so that we get even ripening and we get ripening early enough to avoid rains. Instead, we made the decision to leave more fruit on the vines. It gave the vine something to do instead of ripening a smaller amount faster, which would put the harvest during warm parts of the year. It was a good strategy. We retained acidity, and the fruit came off at the right time."

For the Willamette Valley, the warming has so far been largely for the better. If anything, the weather may previously have been too cool, unable to consistently provide enough heat before summer gave way to fall. "The classic, old-style Oregon vintage, about four out of ten of them would be rain affected," said Peterson-Nedry. "The harvest would be pretty much there, and ripening wouldn't go any further because it would be rained out."

After lunch, Peterson-Nedry took me for a drive to his vineyards. The Willamette Valley is roughly bowl shaped, a break between the coastal mountain range and the upward sweep of the Cascades. Its western stretches lie in the shadow of the mountains and are swept by ocean breezes. They are colder and rainier than the rest. When Peterson-Nedry arrived in 1980, there were no vineyards to the west of his property. It was considered too cool to plant grapes. For the same reason, nobody was planting above seven hundred feet. There was no way to get the fruit ripe enough.

When we reached his estate, Peterson-Nedry pulled his Audi sedan onto the grass next to his vines. "This was our original planting from this block, the first vines we planted here on Ribbon Ridge in 1982," he said. "You can see they have perfect canopies for pinot noir. One or two layers of leaves. They're not excessively vigorous, which you don't want. You need enough to ripen, but no more." I didn't have to roll down my windows to tell it was cooler here than in Napa Valley, where the harvest was already beginning. Peterson-Nedry's grapes were still green. "This is as high as we felt compelled to plant initially, basically five hundred feet," he said.

He turned his wheels and began driving up the hill. The grass was pale green and sprinkled with small white flowers, Queen Anne's lace. He eased his car to a stop at the gentle peak, 692 feet. "I remember the day that I was looking at this with a vineyard consultant back in 1980 when we purchased it," he said. "The conviction was from this point over you don't plant it at all. Don't even think about it. But this is going to be planted. Plus, even more significant than going up in elevation, we will also crest the hill and turn down on the northern side." Peterson-Nedry's neighbors were now occasionally planting at a thousand feet or higher. Acres and acres of vines had been planted to the west of his property. "Where it was once too cold to plant twenty-seven years ago is now no longer too cold to plant," he said.

"My point is that adaptations are going on whether

people know why they're doing it or not," he said. "It doesn't happen overnight, and you don't necessarily know what the cause is. You just get a chance to make some tweaks. Does it end up being different than what you would have done twenty-five years ago? I'm in the position where I know what decisions I made twenty-five years ago. And I'm taking different things into consideration."

He circled his car slowly around the hill. "Short term, people in Oregon are smiling," he said. "Good vintages nine out of ten, as opposed to five or six out of ten. A little bit of warming and we're in fat city. But what if we had gone from bucolic days of sitting in our lawn chairs as we wait for our ripening fruit and instead were going in the other direction? We'd probably be saying climate change is pretty damn significant. We'd be alarmed. The fact that we're going from challenging times to less challenging times makes us pretty complacent. Why would we want to change anything that in the short term is beneficial? And the answer to that is because in the medium term to long term it's not going to be."

So far, climate change has been on the whole good for the world's wine drinkers. A study led by Gregory Jones, a climatologist at Southern Oregon University and the son of a winegrower, tracked the impact of rising temperatures between 1950 and 1999. As a measure of quality, Jones used ratings by the auction house Sotheby's, which rates wines on a sliding scale, from a perfect one hundred to an abysmal forty. Ratings, Jones noted, also reflect revenue. In 1995, for instance, a ten-point rise in the rating of a Napa Valley wine more than tripled its selling price.

In most cases, the vintages had improved with rising temperatures. In the twenty-seven wine regions Jones examined, temperatures had risen on average 2.3 degrees Fahrenheit, producing a corresponding increase in the strength of the wines; faster ripening resulted in more sugar for the yeast to ferment. In Napa Valley, for instance, average alcohol concentrations had climbed from 12.5 percent to 14.8 percent over the past thirty years. In the French region of Alsace, alcohol levels jumped 2.5 percent. Even more striking was the warming's impact on ratings. With few exceptions, they rose dramatically. On average, a 1.8-degree-Fahrenheit rise in temperature yielded a boost of thirteen rating points. The big winners were German wines, with leaps of more than twenty points, and those from the cooler parts of France.

Of course, global warming wasn't the only factor driving up the quality of wine. Improved techniques and technology were undoubtedly helping. Nor was climate change the only reason growers were picking their fruit riper. The last two decades have seen an industry-wide move towards large-scale tastings, in which dozens of wines are compared side by side. "Higher-alcohol wines get better reviews— period," said Randy Dunn, a Napa Valley grower who tries to keep his sugar (and thus alcohol) levels low. "The only way for your wine to stand out among a group of twenty or so is to have maybe 0.3 percent of alcohol more."

The steady climb in alcohol levels also reflects the rise of Robert Parker, the Maryland-based wine critic whose floridly styled newsletter sets the standard for much of the industry. A rating by Parker, who has insured his wine-tasting palate for one million dollars, can make or break a winery. Parker has been credited for the popularization of wine in the United States (asked to name their favorite drink, more Americans list wine than beer) and also stands accused of homogenizing wine tastes. Parker's preference for the heavy, robust, and fruity has walked hand in hand with global warming: growers, eager to please the industry's most influential critic, began to pick their fruit riper; higher temperatures allowed them to do so. "Stylistically, the pinot noir that you're seeing today is different than the pinot noir we grew twenty years ago," said Peterson-Nedry. "Less acid. Big and fleshy fruit, like dollops of jam. All of that ripeness masking some of the spice, some of the nuanced flavors and aromas. Is it different because of Robert Parker? Or is it different because of what we can grow now, how much ripeness we can get into the fruit?" More likely the latter, said Gregory Jones: "You cannot hang fruit until some desired flavor without the climate being ideal. You'd have to bring the fruit in at some point."

The backlash against Parker has already begun. Growers like Dunn and Peterson-Nedry are arguing for a return to lighter wines. If the tide begins to turn, the question will be whether growers will be able to cut strength without dropping quality. Having walked up the alcohol hill in a time of rising temperatures, they may find that it's not so easy getting back down.

KIM NICHOLAS CAHILL PICKED ME UP FROM IN FRONT OF CITY hall in downtown Sonoma. A doctoral student at Stanford University, she was a driving a gray Toyota Prius with an oversized bike rack on its trunk. Our destination was the vineyards at Hunter Farms on the outskirts of the city, where Cahill was studying the effects of rising temperatures on the quality of wine grapes. It was early evening. The sun was low on the vines. Cahill pulled into the gravel lot and parked her car, and we stepped into the vineyards. The ground was loamy underfoot. It was mid-August, early to be picking. Yet the grapes looked ready. They were so dark they were almost black. Wrinkled, slightly deflated, they seemed ready to fall. The harvest was planned for the next day.

We walked down a row, stopping at a cluster marked with a pink ribbon. As Cahill plucked off grapes and slid them into a plastic bag, I broke one off myself and popped it into my mouth. It was very sweet, a surge of flavor across my tongue. Before embarking on this project, Cahill had worked as part of a team that used temperature and rain records to forecast crop yields. From there, it was a short jump to using climate models to predict the impact global warming would have on California's agriculture. Cahill had studied six crops and found wine grapes to be the most robust. Models showed that rising temperatures could devastate the yields of avocados and table grapes and reduce that of almonds and walnuts. Meanwhile, oranges and wine grapes would be less likely to suffer big drops in production.

Yet Cahill wondered if she was getting the whole picture. An overheated vineyard might be able to produce wine grapes, but can it produce good ones? Indeed, growers of fine wine are less interested in quantity than quality. They spend much of their time dropping clusters, cutting back yields in hopes that the remaining grapes will come out better. "Generally wine growing, especially premium-quality wine growing, is not about trying to maximize the crop load," said Cahill. "It's trying to optimize between yields, quality, paying the bills, and making a great wine. Some places don't even care about paying the bills."

I followed Cahill through the vineyard, ducking under trellis wires as she moved from vine to vine. In the laboratory, she would analyze her samples for the chemicals that produced color, flavor, and complexity. She had yet to process all her data, but qualitatively she had a pretty good idea what she would find. Since the middle of the last century, global warming has driven up average temperatures in California by one degree Fahrenheit. Even if we were to suddenly cut back on emissions of carbon dioxide now, the concentration already in the atmosphere is expected to add another degree by 2020. Cahill's early work had found that yields would start dropping if temperatures rose a degree further than that. The impact on quality would almost certainly be felt sooner. "It certainly won't help California to get any warmer," said Cahill. "It's not that we don't have enough time to ripen the grapes. If anything the reverse is true. Unlike the old world, where often warmer vintages are the best, here the cooler vintages tend to be best."

"The average annual temperature in Napa is about five degrees Fahrenheit cooler than in Fresno," she said. "They grow wine in Fresno too. But a ton of cabernet in Napa sells for about four thousand two hundred dollars, and a ton of cabernet in Fresno sells for about two hundred and fifty. There's more than a fifteenfold discrepancy there, and a lot of that is due to climate."

The impact of global warming on the wine industry will be felt first by the grower, and then by the drinker. There's already a lot of variation in today's vintages. Small changes to the climate will at first be discernible only by aficionados of specific estates, who may notice their one-time favorites giving way to what they had thought of as lesser marks. The casual buyer confronted by a wall of bottles may not care that certain blends have become bolder or stronger, or that a one-time table wine has climbed its way to the top shelf.

But meanwhile, individual vineyards will be struggling to adapt. For those who once labored to ripen their grapes, the rise in temperatures will be a blessing. But they will be in the minority. Most wine-growing regions were planted before the world began to warm, and the successful ones were chosen because they produced the best possible wines. When Gregory Jones compared a wine-growing region's yearly ratings with its annual weather data, he was able to pinpoint the best temperature for the grapes within that region. For most of his study period, the areas he had looked at had been slightly cooler than the ideal. By the 1990s, however, climate change had warmed to the point where quality was at or near its peak. As the world continues to heat up, Jones concluded, growers around the globe will likely begin to find that their fruit is ripening too fast. Climate models predict the areas he was studying would gain on average at least another two degrees Fahrenheit by

2049. Growers who don't manage to adapt can expect significant drops in their ratings.

In the early days of climate change, winemakers will experiment with different ways of trellising their plants. Some will try shade cloths or will ramp up irrigation during the hottest days. Wineries that once used machines to concentrate their wine and compensate for rained-out harvests will find themselves considering technology to reduce alcohol levels. Others will start planting in areas they had once considered too cold. "People are going to be a lot more adventurous," said Peterson-Nedry. "They'll go over on north slopes. They'll move into the outside of the valley and go into the cooler Coast Range."

In many cases these changes will come naturally, regardless of whether the grower knows he is responding to global warming. Winemaking is an iterative art. Growers only get a few dozen chances to make a perfect wine in their lifetime, so the best are constantly trying new things—experimenting with new techniques, new varietals, and new patches of their property.

But adaptation can only go so far. Down in the valley below Cain Vineyard, vines are already trellised in the California sprawl. Their tops are bushy. The grapes cluster about halfway up the tall vines. Nothing more can be done with the way the vines are laid out. In the wineries, winemakers have already started to experiment. "Just like in the rest of the world, most wine from California today has been played with," said Jones. "It's been acidified, because it's been allowed to hang too long. Alcohol has been removed or water has been added back. The question is, is there a limit to that kind of tinkering? If you have a sixteen percent alcohol wine and you bring it down to fourteen, does it age as long? Do flavors develop the same way?"

As the mercury climbs, the choices will get harder. As early as 1944, scientists at the University of California, Davis, divided the state's agricultural lands into five climatic regions. Each was assigned a different suite of grape varieties, depending on how much heat it received. Region I, the coolest, was suitable for growing the varieties that ripened fastest: chardonnay and pinot noir. Region II was a bit warmer, also able to ripen merlot and cabernet sauvignon. Region III could handle zinfandel. Region IV was better for port-style wines, and Region V was analogous to North Africa, suitable only for grapes that need a lot of heat.

With climate change, these zones will begin to shift. Jones's research found that a 3.5°F rise in temperature was enough to make a region unsuitable for growing the types of grapes to which it was accustomed. A vineyard growing pinot noir might find it's suddenly better suited for producing sauvignon blanc. A grower banking on chardonnay might discover he's having better luck with syrah. Someone with a field of zinfandel may decide that his land is good only for raisins.

Jones has predicted that within fifty years the best places for Tuscany's famous Chianti wines will be in Germany. The best lands for producing champagne and Bordeaux wines will be in southern England, where a renaissance is already under way. The country makes at least one reputable sparkling wine, the Nyetimber, grown south of London. Already, England probably has more land under wine cultivation than it did during its last heyday, in the twelfth century during the Medieval Warm Period.

In the United States, growers will abandon the hottest climes, exploring the uplands and northern states for new places to plant vines. There will be winners and losers. California is likely to suffer. Oregon will benefit, at least at first. Even those cultivators lucky enough to relocate to areas with the same climate as their old holdings will find their product will have changed; the soil will be different. In an industry that is anything but centralized, resettlement won't take place by fiat. It will happen gradually as growers set themselves up in places once considered marginal that suddenly seem attractive and abandon areas they once relied upon.

Eventually, the impact of global warming will be strong enough to be felt in the cellars. Climate change will make the weather less predictable and subject to wider swings. Growers prepared for a stifling summer may find themselves facing a couple of unusually cool years. Day-by-day variations will wreak havoc in the harvest. Strings of scorching days will stress the vines. Early budding will make vines vulnerable to sudden frosts. A freak, late-summer hailstorm could devastate an otherwise promising season.

In another study, Jones teamed up with colleagues to predict the impact of the next century's warming on the United States' wine industry. Taking into consideration everywhere that premium grapes could conceivably be grown (including areas like the Iowa corn belt), Jones concluded that lands suitable for producing premium wines could shrink by 81 percent. California would be especially devastated. The state produces 90 percent of the country's wine grapes. It's the fourth-largest producer of wine in the world, after France, Italy, and Spain, with global sales of more than $18 billion. Rising temperatures could put much of it off-limits for quality grapes.

Since the quality of what gets turned into fine wine depends so much on the weather, grapes are likely to serve as forerunners for an agricultural industry that will increasingly shift north and upslope. Farmers in the United Kingdom have started to plant olive trees, including an orchard on the north coast of Wales, whose managers say they are planning for a time when the weather just west of Liverpool will resemble that of southern France. In London, gardeners have begun to grow avocado trees. Meanwhile, the center of gravity of the maple syrup industry has shifted from New York and Vermont up into Canada.

Yet while some regions will find themselves happily able to produce new crops, on the whole the impacts are more likely to be negative. Agricultural zones will shift from established regions to zones where the industries are not as well set up. Meanwhile, increasing variability and harsher extremes will pile on the pressure. For the majority of crops, the most important change won't be the simple rise of mercury but the crossing of temperature thresholds. Just as grapes only ripen once the warming exceeds a certain level, other plants might fail to release their pollen if the weather turns too hot.

Meanwhile, winter temperatures may no longer drop low enough to kill pests and parasites. The glassy-winged sharpshooter, a half-inch-long, brown leafhopper, is the major vector for Pierce's disease, a bacterial infection that can devastate a vineyard. Once endemic only in Texas and the southeastern United States, the insect has been working its way up California as the weather warms. "For many of our weeds and insects, their habitable range is defined by how cold it gets in the winter, and that zone is changing," said David Wolfe, a professor of plant ecology at Cornell University.

Water scarcity will form another challenge. As melting snowpack disrupts water supplies, farmers will find themselves competing with towns and cities for water. In Australia, a long drought has sent shivers through the wine industry. When wildfires broke out in the country's national parks in 2003 and 2004, growers feared for their crops. The flames spared the vineyards, but the grapes spent weeks under smoky skies. Drinkers of those years' vintages report a dusty feel and the smell of burning grass.

After wine grapes, the first agricultural products to feel the effects of climate change will be high-value fruit and vegetable crops. "With something like tomatoes or grapes or apples, just one bad day of heat stress can cause deformed fruit," said Wolfe. "Then you've got tons of fruit, but you can't sell it anywhere. A crop like wheat is actually more tolerant to minor shifts in climate to some extent because it's really just focused on total tonnage." Still, for countries that depend on rain for their water, disruptions in precipitation could devastate the harvests needed to feed their populations. Researchers at Stanford predict that southern Africa could lose more than 30 percent of its maize yields by 2030. South Asia could see its rice, millet, and maize crops—staples of the region's diet—cut by 10 percent. "We're leaving the world when you could look at the historical climate for an area and decide what you're going to grow," said Wolfe.

On my way from Napa Valley to the Willamette Valley, I stopped overnight in Elkton, where Terry and Sue Brandborg have been running a winery since 2002. Elkton is a small town in the Umpqua Valley of western Oregon. Three streets wide and six deep, it boasts a well-stocked bakery and a locally run convenience store where a cup of coffee still costs a quarter. The Brandborg Winery is a large metal-sided block, set in a stretch of parking spaces across from the post office and the phone company. In addition to the wine works themselves, it has a large tasting room, which doubles as the town's center for live music, and an upstairs guest quarter where I would spend the night.

I pulled into the parking lot in the late afternoon. Terry was working a small yellow forklift, loading cases of wine into the back of a van. The next day, Sue would drive the delivery. The Brandborgs had invited me for dinner, so Terry and I headed together up to their home. The six-mile drive took us through fields of golden grass. A light mist hung over dark forests hazed in blue. In the passenger seat next to me, Terry smelled of fresh wine and raw wood. The Brandborgs had first come to the Umpqua Valley in 2001, lured by reports that the weather was right for pinot noir. "I went bingo, this is the climate we're looking for," said Terry. "Two weeks later, we came back and found the piece of property that we're heading towards." We left the asphalt and cut up a steep gravel drive. Quail made dashes at the road. "I always like the light in the evening," said Terry.

The Brandborgs' house sits among their vineyards, on a hilltop about a thousand feet above the valley floor. The dining room table is set by two large windows to take in the view. Clouds, lit from below by the setting sun, rippled above the darkening landscape. Big black birds circled slowly beneath us.

"Terry and I still pinch ourselves that we found such a place," said Sue.

"Yeah," said Terry. "Sue and I looked all over California for a couple of years, and quite honestly we were feeling priced out of the market down there. And then I discovered the climate in Elkton, and we came and looked."

Terry wore a gray short-sleeved shirt. He had a thick torso and moved slowly while he cooked. Sue had reddish hair and sharp, fine features. She wore a white blouse with wine-colored dots. The two met at a tasting in Jackson Hole, Wyoming. Terry had already been making pinot noir for several years, from grapes he bought rather than grew. Sue had tasted his wine two years before she ever met him. His was the first glass of red wine she had ever tried.

Now Terry focused on the winery, while Sue tended mostly to the vines.

"I love them," she said. "They're my children, all four thousand seven hundred of them."

"She's named every one," said Terry.

"Well, they're unique, just like people," she said. "People think I'm crazy, but it's true. No two are alike."

"Is there any other crop that's more fastidiously looked after?" said Terry.

During dinner, we drank a Brandborg wine, a syrah, one of the first Terry had produced in Elkton. It was a heavy wine, rich and hearty.

The Brandborgs' vineyards, high in the mountain and cooled by the ocean, are just warm enough to ripen pinot noir. Terry had considered climate change when deciding where to buy. "I knew I wanted to be in a coastal valley," he said. "And I knew I wanted to be on the extreme edge of being able to get grapes ripe." In 2003, when Peterson-Nedry and the growers of the Willamette Valley were struggling to keep their alcohol levels low, the Brandborgs produced wines that took home prizes.

Terry and Sue have all the usual problems of a young estate: dependence on investors and banks, challenges in making their name in the market, being forced to make choices for cash flow rather than the long term. But for the next few years at least, while other growers are waking up to what could be disaster, the Brandborgs will be able to worry about things other than global warming.

"Since our winery is six miles away, we'd like to put a Ferris wheel up here," said Terry. "We'd have a shuttle—our '51 Studebaker pickup, which is down in the barn and in need of major restoration. We could bring people up to the vineyard, stick a glass of wine in their hand, and give them a Ferris wheel ride. And from the top you could damn near see the ocean out that way and that peak out the other way. It's a pretty spectacular view."

"Someday someone's going to say, 'I have a Ferris wheel I'm going to give you,'" said Sue. "I'm holding out for that."

"We'll see," said Terry. "We have other things to do with our money until then, like plant another forty-five to fifty acres."

"We can still dream about it," said Sue.

The sun had gone down. Nightfall had turned the windows into mirrors. The wine was finished.

"Yeah," said Terry. "We can still dream about it."

6

"EVERYTHING IS LATE IN CHURCHILL"

THE ARCTIC, MELTING ICE, AND THE NEW LAND GRAB

There are no roads to Churchill, Manitoba. Visitors and goods arrive by air, rail, or sea. If a resident's truck breaks down, it gets loaded onto a freight car and shipped south to the dealer. The town may lie below the Arctic Circle, but its landscape is nonetheless very northern. The summer sky is powder blue at the horizon. The terrain is low and rocky, broken hills of glacially deposited rock. The wind blows from the north, frosting the streets with dust and spreading a chill not even the summer sun can dispel.

Built on a spit of granite between the Churchill River and the Hudson Bay, the town is Canada's only port on the Arctic Ocean. In the summer, it is flanked by water. In the winter, by ice. "The town itself is a shock to the newcomer," wrote Angus and Bernice MacIver in *Churchill on Hudson Bay*, their otherwise boosterish history of the town. "It appears gray. Then realization comes that this is because it is built on gravel, and the little grass and plant growth around some houses are the result of painstaking toil." With the exception of the airport and a far-flung research station, the entire community—a scattering of small houses around a cluster of hotels—lies close to the train station. While there's no place that can't be reached by foot, wanderers are warned to keep their eyes open for polar bears and to listen for the siren that signals one of the giant carnivores is prowling the town.

By dint of its geography, Churchill may be the best place in the world to spot the great white animals. The community bills itself as the "Polar Bear Capital of the World," and indeed during certain times of the year, the ursine population may outnumber the town's 920 registered residents. Over the years, the bears have learned that the fresh water in the river is the first to freeze, that the bay's currents pack the ice against the rocky coast, and that the waters are rich in ringed and bearded seals. Peak bear season arrives in the late fall, after the snows have come, and Churchill's tour operators crank up their tundra buggies, huge lumbering vehicles with tractor-sized wheels. During these weeks, the bears are thin and hungry. They've spent the summer in a walking hibernation, foraging through the tundra for berries, kelp, peat, and the occasional sluggish goose. They gather at the coast, waiting for the ice to form so their hunt can begin.

But Churchill is changing. The ice in the bay is forming later and breaking up earlier. The bears are becoming fewer. In the past ten years, global warming has lengthened the ice-free season by roughly three weeks, and the polar bear population around Churchill has dropped from around 1,200 animals to roughly 940. "A polar bear loses a kilo [2.2 pounds] of weight every day that he doesn't eat," said Robert Buchanan, president of Polar Bears International, a conservation group. "Not every week. Every day.

Those three weeks are twenty-one kilos. If you're a large male, you can survive that. If you're a cub, you can't. If you're a subadult, you can't. You become weaker, and you're not as fast."

Female polar bears stop reproducing when their weight drops below around four hundred pounds. "Twenty-three years ago when I first went up there, you'd probably see one in seven mothers with three cubs," said Buchanan. "I haven't seen a mother with triplets in five years, and I'm up there a lot. Now what we're seeing is a female with two cubs. And the two cubs are very, very small because she's not able to feed them enough. She's not getting enough food herself." Scientists estimate that by 2050, global warming will strip the world of two-thirds of its polar bears. Those in Churchill will be among the first to go. "Polar bears need ice," said Buchanan. "Without it, they can't hunt, they can't breed, and in most places they can't den. If you're not able to do those things, you won't have polar bears."

Climate change is especially visible in the north, where a few degrees of temperature can change the landscape from frozen to flowing. Surface air temperatures in the Arctic have risen at roughly double the global rate. The ice cover on the top of the world has been shrinking accordingly. Offshore, the impact is obvious—the ice is disappearing— but climate change is being felt all over. In northern Manitoba, communities find themselves suddenly stranded when their winter roads across ice-covered lakes are cut by year-round thaws. Roads, houses, pipelines, and airport runways twist and crack as the once permanently frozen ground melts, sinks, and slips.

The train tracks from Winnipeg to Churchill run seven hundred miles, linking the buckle of Canada's grain belt to its Arctic Ocean port. The ride is meant to take thirty-six hours, a gentle slide through stunted boreal forest. I boarded a train in the evening, just after eight, and crawled into the berth to sleep. I woke up to cloudy skies, a thick gray ceiling. Short, scraggly trees staggered towards the horizon. Meltwater glistened from the gullies besides the tracks.

The view remained the same through that day and through the next. The trees thinned, then thickened, then thinned again. The water pooled and withdrew, pooled and withdrew. Maintenance on the tracks had been patchy, unable to keep up with the heaves of the springtime melt. A couple of freight trains had recently derailed, and the government had lowered the speed limits. With priority being given to the grain trains, we spent long stretches shunted onto the side tracks. When we did move, it was at a crawl—rocking slowly north like a boat in calm waters. By the time we reached the station, we were twelve hours behind schedule.

I had an appointment early the next morning to visit the manager of the port, a man named Lyle Fetterly. The port was not far from my hotel, an easy walk down a short stretch of tarmac, across a set of train tracks strewn with spilled grain. I arrived at 7:30 a.m., when our meeting was set to start. The office doors were locked, and the lights were off. Neither the air nor the sky showed any sign of warmth. The dirt parking lot was hard underfoot. Puddles of brown water trembled in the breeze. The clouds hung low, gray as refrigerator frost.

The tide was out. Fishing boats peered over the edge of the far end of the dock. Two grain ships, long dark hulks, waited for the day's loading. Onshore, the port was dominated by the grain elevator, a huge concrete block flanked by banks of colossal grain silos. The windows, those that weren't broken, were thickened with dust. Dark gray cement patchwork meandered across the otherwise rigid facade, as if Paul Klee had defaced a Mondrian. A red-rust fire escape zigzagged its way towards the top, where a couple of workers, having arrived early, used its highest platform for an early-morning cigarette.

The administrative offices were tucked into a small building in the shadows between the elevator and its silos. The port's employees were arriving by ones and twos, on bicycles or by pickup truck. When Fetterly's secretary pulled in just after eight, she unlocked the offices and led me inside. Glass doors opened onto a small hallway and another set of glass doors. The carpeted floor was tracked with grain dust. On a poster on the wall, a polar bear clutched a stalk of wheat between his teeth, with the caption: "Canadian Wheat—it's good, eh!"

Fetterly's secretary was surprised to learn that I had an appointment, and after calling her boss twice to establish he wasn't answering, she made a third call to ask someone to swing by his house and pick him up. When she was turned down, she reached for her car keys.

"Everything is late in Churchill," said another secretary. "It's like its own time zone."

• • •

Established in 1717 by the Hudson's Bay Company, the quasi-statal corporation that colonized much of Canada, Churchill has spent its history limping from failed industry to missed opportunity. Geographically and economically isolated, the town never quite scraped together a reason for its existence. It was always too far, too marginal, too frozen, too sleepy.

As a sporadically successful trading outpost in a region rich in fur and copper, the town was important enough to be defended with a stone fort, but not important enough to fight for. When the French attacked in 1782, the garrison gave up without a shot. Back under English control after the Treaty of Paris, the town saw its first boomlet in the closing years of the nineteenth century when residents spent their summers hunting beluga whales. At its peak, they were slaughtering enough to feed two refineries and were exporting up to thirty tons of whale oil a year—until American competition took the bottom out of the market.

In 1927, Churchill got another boost when the Canadian government declared that the town's deep-water port would serve as the terminus for the railroad from Winnipeg. It was a transformational moment. The town had always been more of a worksite than a community. The European population was almost entirely male. The local tribes—Cree, Chipewyan, and Inuit—who supplied the traders would come and go. Those who stayed nearby rarely lived within Churchill itself.

The town, which lay on the other side of the river from where the train would arrive, was dismantled—houses, churches, shops—and pulled by sled across the frozen water. Meanwhile, thousands of tons of cement and steel were brought in for the construction, pulled over the icy bay in the dead of winter. "Tractors were used and the men fought their way through temperatures as low as forty degrees below zero, winds of hurricane proportions, and deep snow that at times reduced the progress of the machines to less than a mile an hour," wrote the MacIvers. "In one place a tunnel had to be forced through the drifts. When the drivers stopped during the night it was the task of one man to turn over the motors every hour or second hour depending on the lowness of the temperature."

Churchill was thrown open to settlers. A community hall was built. And then a school. The tracks were laid. The silos were raised. Grain cars rolled in. But the port never took off. Churchill was too far, and the ice-free season was too short, just a few months long. The town remained a seasonal place, emptying almost completely during the darker months. When the Second World War broke out and shipping stopped altogether, nobody really missed it. The next twenty years were the town's most prosperous as it hosted first an American military base and then a rocket research range, operated in turn by the Canadian and United States governments. When the Canadian army finally pulled out in 1964, the town resumed its slink into obscurity. Grain was arriving again, in dribs and drabs, but never in quantities large enough to do much more than sustain the port. Tourism picked up some of the slack, but Churchill's permanent population continued to shrink.

Then, in 1997, the Canadian government sold the port and railroad to OmniTRAX, an American transportation company based in Denver, Colorado. For the politicians in Ottawa, it was an opportunity to privatize some of its most crumbling infrastructure. For the firm, it was a chance to pick up a roughened asset, with what it hoped was unused potential. The tracks cost the company $11 million, according to *Forbes* magazine. The port was thrown in for just seven dollars more and the promise to bring it up to code. With a little luck and a lot of work, the deal would one day pay for itself. Nobody thought much about it— until the ice began to melt.

Fetterly arrived just before ten, explaining that he had been caught in a phone call with his lawyer. He was a young man, not tall, with business school diction and wearing a black cotton sweater. His thick dark hair was cut short on the sides and slightly longer on top. His sideburns dropped to the hinge of his jaw. His office window looked out towards the ships and onto the grain galleries through which wheat shuttled on long conveyor belts. He settled in behind his desk, inviting me to sit across from him, and when I told him I was interested in how climate change was affecting his business (I imagined it would be beneficial), he dropped his head to collect his thoughts.

"It's a touchy subject," he said. "Is global warming going to be good for the Port of Churchill? Yeah, it has the potential to be good." Officials at the port had been keeping records of the depth of the ice in the mouth of the river since the 1930s, and Fetterly had noticed significant declines. "If there was two meters [six and a half feet] fifty years ago in the height of January, there's maybe one and a half meters now [almost five feet]," he said. Global warming had stretched the shipping season by roughly two weeks. The waters were now navigable about four months out of the year. And the change seemed to be accelerating. What was a disaster for the Arctic town's polar bears had the potential to transform its port.

Churchill enjoys several advantages over its competitors. The port is capable of handling ocean-class vessels with none of the locks or shallows of the routes through the Great Lakes. A ship can pull in, load up, and head out without ever leaving deep waters. Distances are also shorter. Vessels make a short slide out of Hudson Bay, cross the Atlantic Ocean just under Greenland, and cut hundreds of miles off the journey to Europe. Compared with the port of Thunder Bay on the Ontario coast of Lake Superior, the distance from Churchill to Liverpool is nearly a quarter less. Ships that sail from Churchill to Oslo cut the mileage by a third.

Yet none of that has been able to make up for the ice. Frozen water doesn't block only ships. A shorter season means less money for infrastructure—the facilities to add volume to the flow of grain, to handle containers or petroleum. A dollar of investment in a year-round port yields returns twelve months out of the year. In Churchill, that same dollar works only in the summer months. Each additional day without ice means more volume, more money, more interest from shippers, and better returns on investment. "Does this mean the Port of Churchill wants to see global warming?" said Fetterly. "No. Of course not. Nobody does. But we can guess at what the results will be. Would it be responsible for us to harness that new reality? It would be stupid not to."

Since OmniTRAX made its purchase, climate change had lengthened the shipping season by two weeks. The value of the port's infrastructure—in terms of how much work it can handle—had increased by nearly 15 percent. That summer, the port would ship 621 metric tons of grain, the most the port had seen in thirty years. The following year's volumes were expected to exceed a million metric tons. And news was getting out. Churchill also handled its first domestic delivery—a load of Durham wheat that was sent to Halifax—and its first inbound batch of Russian fertilizer, shipped in from Murmansk. "There's a limited amount of business the Port of Churchill can do in a day, be it shipping grain, receiving grain, transloading cargo, or importing cargo," said Fetterly. "What is global warming going to do to Churchill? It means we have capacity there to grow."

The first Europeans to enter the Churchill River, a group of sixty-five Danish explorers seeking shelter from an autumn tempest, arrived in September 1619. The expedition was led by Jens Munck, one of the top officers in the Danish-Norwegian navy. He had been given two ships and charged with finding the Northwest Passage, the holy grail of Arctic exploration. Sailors had been seeking the shortcut to Japan and China through the straits of the Arctic since 1497, when the Venetian explorer John Cabot had mistaken the North American continent's eastern coast for the shores of Asia. When the Danes blew into the river, the

Hudson Bay seemed the most likely place the route would be found.

The storm lasted four days, and by the time it was over, Munck had decided to stay until spring. The river provided a large natural harbor. It seemed safer to risk months on the ice than to attempt a return so late in the season. But winter in the New World proved harsher than anything Munck had experienced. Fresh meat was scarce, and his crew quickly sickened. Two were dead before Christmas, including one of the ship's surgeons. By January, the pastor and the other ship's surgeon were too weak to leave their beds. "The illness that had fallen upon us was rare and extraordinary, with the most peculiar symptoms," Munck wrote in his diaries. "The limbs and joints were miserably drawn together, and there were great pains in the loins as if a thousand knives had been thrust there. At the same time the body was discolored as when someone has a black eye, and all the limbs were powerless."

Munck's men succumbed one after the other. With the ground frozen, the weakened survivors found it impossible to bury the dead. By April, Munck was alone, surrounded by corpses, and barely strong enough to walk. He struggled from his cabin and onto the banks of the river, where he was surprised to find two others from his crew. "The three lived onshore under a bush in front of which they managed to have a fire part of the time," wrote the MacIvers. "When growth started the men crawled to every plant that showed a bit of green, clawed it up, and sucked the root; their gums were so sore and their teeth so loose that to chew was impossible." The arrival of spring opened the waters, and they could begin to fish. Birds returned from the south. Munck and his two companions were the expedition's only survivors. Somehow, they managed to lower their smallest ship back into the river and sail it back to Europe.

Munck's travails are typical of a history of exploration in which the protagonists regularly risked their lives. The Arctic is one of the most inhospitable places on Earth, a frozen desert where even the process of decomposition slows to a crawl. The men who sailed their ships into its frigid waters in search of the Northwest Passage did so for reasons that ranged from the commercial to the gallant. In 1611, when a mutinous crew cast Henry Hudson adrift in a small boat with his teenage son and seven loyal crewmen, the Northwest Passage was imagined to be a route to riches. By the time John Franklin and the 128 members of his expedition froze to death in the Canadian Arctic in 1845, the search for the passage had been cast as a quest for glory and scientific knowledge. What didn't change was the ice and the danger.

"The cold brought frostbite and amputation, numbing headaches, and stupor to overwintering ships' crews," wrote Barry Lopez in *Arctic Dreams.* "No kind of clothing or shelter could keep it entirely at bay. The cold made the touch of metal burn and all tasks more difficult, more complicated. Even to make water to drink was a struggle. . . . In spring the light came. It gave men 'an extravagant sense of undefined relief,' and in their innocence and abandon they became snowblind. Their eyes felt as if they rested on needles in sockets filled with sand. In harness they dragged sledges across the trench and rubble of sea ice and through vast sumps of soft snow. Consumed by the immenseness of the land, men tramped on mindlessly and fell over dead—of exhaustion, of fatal despair or miscalculation. Died in a tidal crack that suddenly opened, or from a ridiculously simple accident. Starving men ate their dogs, and then their clothing, and then they turned to each other."

The first successful sailing of the Northwest Passage was embarked upon only in 1903, one year before the United States broke ground on the Panama Canal. Explorers had mapped the eastern and western Arctic coasts. All that was left to do was connect the dots. The Norwegian explorer Roald Amundsen entered the Canadian Arctic through Baffin Bay in the east, spent two winters iced up in Nunavut, and emerged off the coast of Alaska. Frozen in once again, he sledged five hundred miles south to the nearest telegraph station to announce his triumph. Nobody was surprised that the trip had taken three years. It had long been clear that the passage was too perilous to be practical.

I had timed my visit to Churchill to coincide with the arrival of the CCGS *Amundsen*, a Canadian research ice-breaker named for the Norwegian explorer. It was just before nine in the evening, and the sun clung patiently in the sky as I was met at the docks by Randy Spence, the head of technical service and security for the Port of Churchill. Spence was a quiet man, stocky without being big, with a curved black mustache and a black leather jacket. Spence and I drove to the end of the docks to wait. We sat in the truck with the windows up. In the river below, beluga whales cut through the sea surface, roiling through the water like the coils of a serpent. Flocks of birds, already considering the trip south, spun in the air. The port sits just inside the mouth of the Churchill River, protected from Hudson Bay by the low ridge on which the town sits. We wouldn't see the ship until it was almost upon us.

Spence was receiving regular updates on his radio, and when the *Amundsen* was about twenty minutes away, we were joined on the dock by another truck. We stepped out to greet the new arrivals, a clutch of graduate students, professors, and researchers who had flown up to greet the ship, switch places with the scientists on board, or check up on their experiments. Among them was Gary Stern, a researcher with Canada's Department of Oceans and a professor at the University of Manitoba, who had been the chief scientist aboard the *Amundsen* the year before when the icebreaker ran the Northwest Passage and found it largely free of ice.

In late October, the ship had entered the Fury and Hecla Strait, the normally ice-clogged, half-mile-wide break between mainland Canada and Baffin Island. Five centuries' worth of explorers would have been amazed. The water was flowing freely. "There was no ice," said Stern. "None. The *Amundsen* was the first ship that ever went through that strait that late in the season. Ever. Even in summertime, it's very difficult for ships to go through there." On its way out of the strait, the ship stopped at a fishing village. "The mayor was saying it was incredibly difficult for them," said Stern. "They generally do their hunting for caribou on Baffin Island at that time of year. They usually go by snowmobile because there's good enough ice there.

But they couldn't that year. They actually had to charter a plane to fly over there."

The *Amundsen* is big and red, with a bright white smoke-stack shaped like a dorsal fin on which a giant crimson maple leaf stands out in sharp relief. When it finally rounded the corner, it was impossible to miss against the dusk-washed river. Formerly a Canadian Coast Guard icebreaker, the ship was bought from the federal government for a dollar by Canadian scientists, who then spent $27 million refitting it as a science vessel. The ship is 320 feet long, crewed by forty members of the Canadian Coast Guard and staffed by forty scientists. It has space for a helicopter, four cranes, and three launches including a swamp skimmer similar to the type used in the Florida Everglades. Thrusters at its stern and sides allow it to steer in any direction. As we stood shivering on the shore, the captain executed a complete U-turn, sliding into the dock in a slow-motion aquatic skid.

Churchill was the site of the first crew change in what would be a fifteen-month trip from Montreal across the Canadian Arctic and back. The lead scientist on the first leg was David Barber, a professor of environment, earth, and resources at the University of Manitoba and one of the world's leading experts on sea ice. A big, thickly bearded man with curly hair streaked in black and white, he looked like an Arctic explorer from central casting. He had a large, square face and matte blue eyes and wore a blue jacket over a light khaki shirt and Birkenstock sandals over blue socks.

As the students and scientists on board spilled off the ships and the crew began the loading and unloading of the gear and supplies, Barber took me for a brief tour of the deck, explaining each of the ship's major instruments. A ring of bottles about as thick and long as a man's arm could be sunk four miles into the ocean and opened remotely to collect samples of the water. A deep-water shovel scooped up sediment. A bank of nets with computer-controlled openings sampled the sea life. "In the physical world, we like to say that we study everything from the bottom of the ocean to the top of the atmosphere," said Barber. "And in the biological world, we study everything from viruses and bacteria right up through the different food chain elements to whales and people. The idea is to connect those two things together."

Barber stopped at his office. Furnished with a wood-paneled desk, a small table, and a round three-person couch upholstered in a green striped fabric, it looked like it could be an office anywhere, until I looked out the window and saw the river skipping past. Barber was handing the project over to a team of doctors, nurses, and scientists who would spend the next six weeks conducting a health survey on the Nunavut coast. His bags were in the next room. "I used to be a climate change skeptic," he said. "I figured that this used to be a part of the natural variability and the natural cycle, until about ten years ago, when it really started to dawn on me that we're headed into a very strong trend downwards." Barber's conversion began in 1991 when Mount Pinatubo in the Philippines erupted, throwing dust and ash deep into the stratosphere. The aerosol particles formed a haze, dimming the sun. Global temperatures dropped briefly by one degree Fahrenheit. "I thought, if that happens with dust particles, why can't we be doing it with gas particles?" said Barber.

"I started paying more attention to what was going on with temperatures in the Arctic," he said. "We started to see changes in the ice. We started to see thaw holes in the bottom of the ice. It was actually melting from underneath and not from on top. We started talking to the Inuit about this, and they said, in the region they were in, 'This never happens. We don't know what's going on. It's like the ocean is warmer.' For them climate change is a very real thing. They've even started to develop new words for things they used to never have before, like 'sunburn' and 'bumblebee.'"

"We started to do fall projects where we couldn't work on the ice, because it just wouldn't form," he said. "We couldn't get out onto it. I started having to do things like build and design special boats that would allow us to get out on this ice when we used to use snow machines. The evidence just bombards you."

In September 2007, the European Space Agency announced that the Northwest Passage was fully navigable for the first time since records began. Since Roald Amundsen's successful crossing, 110 boats have successfully made it through the passage. Eighty were icebreakers or commercial ships with hardened hulls. But as the ice has receded, recreational ships have started to try their luck. "There was hardly any ice," Roger Swanson, a seventy-six-year-old Minnesota pig farmer turned yachtsman, told the *Wall Street*

Journal as he finished the trip that year. He had tried twice before, in 1994 and 2005, but had been turned back when the passage froze. "It has been a beautiful trip," he said.

The North Pole's ice grows and recedes with the seasons. In the darkness of winter, it fills in the Arctic Ocean, pushes up against northern Russia, slides down the coasts of Greenland, and stretches tentacles into the waterways of the Canadian Archipelago. It seals up Hudson Bay and pushes through the Bering Strait all the way to Siberia. Under the summer sun, it forms a roundish cap, clinging to Greenland and northern Canada. Scientists track changes in the ice by measuring the amount left in the fall at the end of the melt season, when the sun starts to slip away. In 2007, the ice cover reached a new minimum, a 23 percent drop from the previous record in 2005. A stretch of frozen white the size of Alaska, Texas, and California was suddenly running with waves.

If the ice cover continues to shrink at its historic rate, summers at the top of the world will be ice free all the way to the North Pole by 2050. It's much more likely that the ice will disappear far faster, as breaking floes accelerate the warming of the Arctic. "You have black ocean covered with a white surface," said Barber. "So when you get a lot of sunlight in the summertime, that white surface reflects the light back into space again. Remove that ice cover, and you've got a black surface, and it absorbs like crazy. All this energy coming in from the sun that used to be reflected to space is now being absorbed by the ocean."

"Climate change is really changing the Arctic from an environment that used to have multiyear sea ice in the center, which is ice that survived a summer and went on the next winter to grow again," said Barber. "It's getting rid of that kind of ice and replacing it with first-year sea ice." The difference between first-year ice and the older ice of the central Arctic is the difference between limestone and marble. New ice never thickens more than two yards. Multiyear ice can grow to be eight yards thick. Leached of salt, it can be as hard as concrete. "First-year sea ice is much easier to work with," said Barber. "It's softer. It's thinner. It's more pliable. You can design icebreakers or drill ships to withstand first-year sea ice quite simply."

"So when you say you have a seasonally ice-free Arctic, what that really means is you no longer have multiyear sea ice," Barber said. "It means shipping throughout the year will be very possible."

PERCHED ON NORWAY'S NORTHERN COAST, WHERE THE SCANDI-navian peninsula shatters like an ice floe into the Arctic Ocean, Hammerfest claims to be the northernmost city in the world. The sun disappears for two months out of the year, but the warm waters of the Gulf Stream keep its deepwater harbor free of ice. It was mid-February when I visited. The sun was out. The hills outside of town were covered with snow, but the temperature was not far below freezing.

Once a regional center of commerce, a frozen frontier town booming on whale oil, seal skins, and fish, Hammer-fest was the first northern European city to install electric streetlights. But for most of the last half century, it stag-nated. Trade favored warmer climes. Its largest industry, a frozen salmon factory, was cutting jobs. A large reserve of natural gas had been discovered ninety miles offshore, but the water was considered too deep and the markets too distant for it to be extracted profitably. A city in decline dies through the ambition of its young; year after year, they moved away and stayed away. "People weren't eager to build houses, or even paint them," Arvid Jensen, a long-time resident, told me. "You could see it in people's eyes; there was no optimism."

Then, in 2002, with gas prices rising, the Norwegian state oil company Statoil declared it had found a way to exploit the reserves. Gas would flow through an underwater pipeline from deep-sea wells on the ocean floor to an onshore processing plant. It would be cooled, liquefied, loaded onto tankers, and shipped to the United States. The project was named Snow White, and for the residents of Hammerfest, it was a fairy-tale ending. When city employees compiled population figures three months after the announcement, they found that people had stopped moving away. "This is just the beginning," said Jensen, who now chairs Petro Arctic, a trade group of 350 northern Norwegian companies serving the oil and gas industry. "The Barents Sea is an enormous sea, and we have just started to work."

The ice isn't the only thing that is thawing in the north. Climate change has unlocked a geopolitical stasis that had lasted since the age of exploration, through two world wars and the Cold War. The Arctic is rich in metal, minerals, and petroleum. The early administrators of the trading post at Churchill were lured as much by the copper ore flowing down the river as the fur and whale oil. Gold discovered along the Klondike River in the Yukon in the late nineteenth century made the region synonymous with speculation, greed, and sudden wealth. More recently, Canada has discovered diamonds. In the five years after the country opened its first pits in 1998, miners unearthed nearly $3 billion in diamonds, roughly a three-pound sack of the precious gemstones every day. The country has become the world's third-largest diamond producer in terms of value, after Russia and Botswana. And there's more to be found. The United States Geological Survey estimates that the Arctic contains roughly a quarter of the world's undiscovered reserves of oil and natural gas.

When it comes to resources, "the question has always been not whether or not they're there, but at what point do they become economically viable to exploit," said Rob Huebert, a professor of political science at the University of Calgary. "You start getting into an Arctic that's more maritime accessible. You start combining that with a hundred dollars a barrel. Even some marginal sources start looking pretty good. People are going, Holy man! Joining the dots."

The stakes are high. Norway's latest gas and oil finds lie above the Arctic Circle, at latitudes that would be frozen if not for the Gulf Stream. The country's petroleum wealth has allowed it to maintain one of the most generous welfare states in the world. The sick get free health care and up to a year of paid leave. When a woman gives birth, she can choose between taking off ten months at full pay or a full year at 80 percent of her salary. Students are not only exempt from tuition fees, they are automatically eligible for loans each year, nearly a third of which gets written off on graduation. Prices are high, but so are wages. A pint of beer might cost nine euros, but the graveyard shift at a late-night deli in Oslo rakes in twenty-two euros an hour.

In Stavanger, the country's oil capital, unemployment runs at 1.5 percent. At a university job fair for petroleum companies, company reps approach students timidly like overdressed boys at their first dance. A French technical university offers potential recruits grapes and chocolate. The engineering union is making waffles. An oil contractor called Fabricom is raffling an iPod. "Every second week there's another company coming to buy pizza and beer for the students," said Tonje Bye Lindvik, a thirty-year-old engineering student. In Hammerfest, where the annual municipal budget is sixty-two million euros, the Snow White project was expected to generate more than twelve million euros in yearly property taxes. When I visited, the municipality had already launched a fifty-million-euro renovation of the schools and started construction on a twenty-million-euro Arctic Cultural Centre and a twelve-million-euro dock. The year before, the city ringed its hills with a brand-new fence, to keep out a uniquely Arctic pest: "We have a problem with reindeer here," said Bjørn Wallsøe, a city worker. Even before the gas began to flow, several years of revenue had already been spent.

As long as the Arctic was frozen and its resources lay safely locked up out of reach, the neighboring countries were able to agree to disagree on who owned exactly what. Now, the rapidly opening region may hold the world's highest concentration of contested areas. The Arctic is the one part of the world where Norway, home of the Nobel Peace Prize, shows more muscle than smiles. The government describes the high north as its top strategic priority. The country has a long-standing dispute with Russia over exactly where they should draw their ocean border. For now, the two sides have agreed on a gray area that is off-limits to fishing and oil exploration. Norway has claimed the Svalbard archipelago, a sparsely inhabited cluster of islands far north of the Arctic circle, since 1925. That Moscow has never recognized Oslo's claim to territorial rights extending 230 miles from the island coasts has not stopped the Norwegian Coast Guard from boarding Russian ships it says have entered its waters.

Rising temperatures in the north have led Canada and Denmark into buffoonish chest beating over Hans Island, a clump of icy rock the size of a football field that lies between Canada's Ellesmere Island and the west coast of Greenland. In 2005, Canada sent its defense minister, Bill Graham, to lace up his hiking boots and take a walk on the island. The two countries had taken turns landing sailors. In 2003 and 2004 Copenhagen had sent warships to erect the Danish flag. The Canadians were responding by erecting a plaque, the Maple Leaf, and an Inuit stone marker. Both sides took out ads on Google, and the conflict nearly reached the "freedom fries" level when the Canadian parliament jokingly threatened to ban Danish pastries. "I wasn't there to make some big dramatic statement," Graham told the Canadian news agency when he returned from Hans Island. "My act of going there was totally consistent with the fact that Canada has always regarded this island as a part of Canada. . . . I was just visiting Hans Island the way I visited other facilities of Canada's."

Behind the scuffle lay a concept of sovereignty that hasn't been much updated since the days when explorers would plant a flag and claim a land for king and country. Disputants may be able to choose to resolve conflicts in the international courts, but the cases are likely to turn on arguments that have changed little since the fifteenth century. Land belongs to whoever has had a physical presence, the country that historically has been able to exert control, and that is likely to prove the deciding factor.

In the run-up to the 2006 Canadian elections, Prime Minister Stephen Harper pledged to defend Canada's sovereignty in the Arctic. The next year, he announced he would be building eight new Arctic patrol ships and a new army training center in Resolute Bay, north of Baffin Island. Nearby, just inside the eastern entrance to the Northwest Passage, a deepwater port in Nanisivik would be refurbished for docking and refueling. "Canada has a choice when it comes to defending our sovereignty over the Arctic," he said. "We can either use it or lose it. And make no mistake, this government intends to use it."

Canada's most important clash is not with Denmark over Hans Island, but with the United States. As with Russia and Norway in the Barents Sea, the two countries are at odds over how to draw the border between Alaska and Yukon into the water. More pressing, given the speed with which the ice is melting, is a disagreement over the status of the Northwest Passage, which Canada claims as internal waters and the United States argues is an international strait.

The treaty under which Arctic countries will most likely press their claims is the United Nations Convention on the Law of the Sea, an international agreement on how to divide up, manage, and protect the world's oceans, which the United States has not ratified. Washington argues that the passage fits the legal definition of an international strait, which would prohibit Canada from blocking shipping. Canada argues that provisions for ice-covered waters will allow it to impose safety and environmental regulations. The Inuit, Ottawa points out, have long used the frozen straits the way other nations use land. Beginning in the 1960s, the United States has sent three ships through the passage to test Canada's claims. After the most recent, a 1985 trip by the USCGC icebreaker *Polar Sea*, the two sides came to an agreement. The United States would stop sending icebreakers without its northern neighbor's consent if Canada would agree never to withhold it.

As long as ice clogged the passage, the point was not considered worth pressing. But now that it seems increasingly viable, Washington is holding up the three trips as proof the passage was historically used as an international strait. Canada answers that three is not enough. "It's almost like, How many angels can dance on the head of a pin?" said Huebert. "How many nonpermissions make it an international strait? International law is very unclear on this. What happens if the ice melts for half the year? Two-thirds of the year? Do you count when you had the ice cover in terms of when the agreement was reached?"

Two weeks before I arrived in Churchill, Russia had raised the stakes when one of its nuclear-powered icebreakers dropped two minisubmarines through the ice above the

North Pole. The two crafts dove more than two and a half miles under the sea to plant a rustproof flag made of titanium metal into the gravelly ocean floor. The Law of the Sea allows a signatory to claim underwater resources nearly seven hundred miles offshore if it can prove the underlying seabed is an extension of its continental shelf. The Russian expedition was part of a mapping effort to establish claims as far as the North Pole.

Combined with the news of the melting ice, it was a Sputnik moment. Later that month, the United States Coast Guard sent its newest and most technologically advanced icebreaker, the USCGC *Healy,* on a four-week research trip off the coast of Alaska. Two months later, the coast guard announced it would be opening its first Arctic operations base near the country's northernmost point. That same month, the Law of the Sea Treaty passed the Senate Foreign Relations Committee. President George W. Bush had argued that it would "give the United States a seat at the table when the rights that are vital to our interests are debated and interpreted." Unratified, the treaty was of no use to the United States in pressing its claims. Conservative senators, concerned that it would impinge on the nation's sovereignty, had kept it on ice for twenty years. It took a changing climate to work it loose.

One afternoon in Hammerfest, I watched the speck of a plane thread a contrail through the sky. The twin tracks started at the horizon and stretched straight up. The jet was on a polar route, flying from somewhere in Asia to somewhere in North America across the shortest possible distance. Ships that take the Northwest Passage from Europe to Asia will cut more than four thousand miles off the trip through the Panama Canal. They won't have to go through locks, and ships of all sizes will be able to pass through. It won't be long before commercial vessels decide to test its waters and Canada's claims to sovereignty.

The environment in the high north is particularly fragile. Arctic ecosystems develop on time frames that are nearly geologic in scale. "In the short term, uncertainties about the weather, the availability of search and rescue, and the movement of multiyear ice will—along with higher insurance premiums—dissuade reputable companies," wrote Michael Byers, a political scientist at the University of British Columbia, in the *Toronto Star.* "But less reputable ones might take the risk. There are quite a few rusting-out tankers with Liberian flags and disgruntled creditors sailing on the world's oceans. International shipping in the Arctic carries with it serious environmental risks. An oil spill would cause catastrophic damage."

As the ice melts, the United States and Canada will inherit whole new stretches of coast to monitor. "The big problem in the north is that our radar systems, our satellite coverage, and our ability to see through the population centers themselves are much less than on the east or west coast," said Huebert. "To a certain degree this whole issue of sovereignty is almost a moot point. The question is do you actually have the capability of being able to know someone is in your waters, and second of all to do something about it?"

"The hottest part in the world is the cold spot," said Lloyd Axworthy, a former Canadian foreign minister. It was under Axworthy's watch that the government sold the Port of Churchill to OmniTRAX, and the town had responded by naming the short stretch of road to the port after him. "There are some degrees of opportunity in climate change, but in a typically human fashion, we're approaching it with flags flying and gunboats landing," he said. "It's an absurd situation where you've got Russians, Americans, Canadians, Danes all planting flags on undersea shelves and little spits of islands, firing up the frigates. We're back in the nineteenth century. If it wasn't so serious, you'd have to laugh at these sorts of antediluvian characters who are running these countries."

"You've got a Russian regime that's retreating into extreme nationalism. You've got an American that's been playing the same game. If what you're doing there is putting more military ships, and more military bases, and sending more rangers, more guys with flags, at what point do they bump into each other?"

"When I was foreign minister, I got a call one night when I was at a street party in my constituency," he said. "It was from Madeleine Albright." A group of fishermen in Prince Rupert Harbour, accusing Americans of fishing in Canadian waters, had surrounded an Alaskan state ferry and were preventing more than three hundred passengers from leaving. "She said, 'If it wasn't for you, Lloyd, the marines would be coming in,'" said Axworthy. "We kind of laughed. But the reality is that there's a lot of things up there that you're not going to be able to control."

"The Arctic for a long time was an interesting area," he said. "The Russians and Americans played sort of tag under the ice. Is there going to be a nuclear exchange? No. Are we going to be Palestine and Israel? No. Do I predict that you'd have U.S. Marines shooting at Inuit rangers? I couldn't imagine it."

"But throw Russians into that mix?" he said. "Hmmm, who knows?"

7

"AN ELEMENTAL KIND OF EXISTENTIAL THREAT"
SOUTH ASIA, DISAPPEARING GLACIERS, AND REGIONAL CATASTROPHE

Viewed from above, the Brahmaputra looks less like one river than many. A tangle of channels weaves a watery braid through the South Asian floodplain, splitting and looping around diamond-shaped sandbars. The river's spread can stretch as much as fifteen miles, but a person crossing in the dry season will spend more time walking across island fields than paddling against currents. "It's a big river and an unpredictable one," said Manoj Talukdar, an economist at Cotton College in Guwahati, the capital of the Indian state of Assam.

From its origins in the western Himalayas, the Brahmaputra cuts nine hundred miles east through the South Tibet Valley, breaking south just before Burma to cascade out of the mountains. By the time it has reached the flatlands of northeastern India, the river has tumbled more than three miles in elevation. In the remainder of its run through India and Bangladesh, it travels a distance equivalent to that between Boston and Miami and drops just five hundred feet. Flat, powerful, and slow moving—heavy with silt, sand, and soil—the Brahmaputra sweeps the Tibetan highlands towards the Bay of Bengal.

The result is a river that flows as much with earth as with water. With every flood, it redraws its course: in one place, a stretch of bank disappears; in another, an island is born. "Land cannot be lost," said Talukdar, who has been advising the Assamese government on erosion control. "It has to deposit elsewhere. And what's happening is that it's going further downstream. The local people are losing their land, and the land is coming up elsewhere."

The river's islands offer some of the richest ground in the region, with soils replenished yearly as the subsiding flood deposits its sand and silt. Yet, life on the river is nothing if not precarious. A shift in currents might cut at one bank while new deposits add acreage to a neighbor's farms. The island fields may be fertile, but they are also frequently flooded. "We first put the food and clothes on the bed, when the water rise is not too much," one Bangladeshi family told the Indian journalist Sanjoy Hazarika. "Then when it goes above the level of the bed, we put tables and planks together and live on top of that, or finally we go to the roof of the hut—with our families, goats, and chickens."

The geology of the river's islands makes them favorable for more than agriculture. Isolated, ephemeral, and insecure, they make an ideal point of passage for a clandestine search for a better life. Their shifting shapes make it difficult to accurately demarcate the border, never mind patrol it. And once across, it's easy for new arrivals to slip into communities far from the scrutiny of the state. Particularly near India's border with Bangladesh, they form a crucial transit point for an exodus from one of the world's poorest countries. So even as the soil slips south, a countercurrent of illegal immigration pushes north.

Both flows—erosion from upstream and the movement of people from the river's lower stretches—are likely to increase as the climate changes. The Himalayan glaciers have already started to melt, in some places retreating by dozens of yards every year. The Intergovernmental Panel on Climate Change estimates that if the world continues to warm at its current rate, the last of the ice will be gone as early as 2035. The glaciers have been melting from their sides, withdrawing uphill as they surrender their water to the seasonal flows. Meanwhile, they are also thawing from the top, forming large lakes held back by dams of ice or glacial debris. When these break, the unleashed torrents rip through fields, burst dams, and wipe out bridges for up to 120 miles.

Meanwhile, downstream where the Brahmaputra meets the Ganges, the country of Bangladesh may be the one place where climate change will be hardest felt. A population half the size of that of the United States crams into an area a little smaller than Louisiana. The vast majority of its inhabitants live on the world's economic margins, where the slightest blow can mean destitution. With floodplains stretching across roughly 70 percent of its territory, Bangladesh is uniquely vulnerable to flooding, rising seas, and storm surges. Most of the country lies within six yards of sea level.

The World Bank estimates that a one-and-a-half-yard rise in the level of the oceans would flood 18 percent of its territory. In the Sundarbans National Park in southwest

Bangladesh, the trees have begun to die; rising seas have turned the groundwater salty. In a normal decade, the country will experience one major flood. In the last ten years, the rivers have leapt their banks three times, most recently in 2007, when eight million people were affected. That winter, Cyclone Sidr, a category four storm, tore into the country's coast. Twenty-foot waves flattened tin shacks, ripped through paddies and shrimp farms, and plunged the capital into darkness. The Red Crescent estimates that as many as ten thousand people may have died. "Bangladesh has always had to live with widespread floods," Sabihuddin Ahmed, a former permanent secretary at the country's Ministry of Environment, wrote in London's *Independent* newspaper. "With climate change the temporary flooding we see during the wet season is becoming permanent," he continued. "That might not be such a problem were Bangladesh not so densely populated. . . . When these people's homes and crops are flooded forever, where will they go? What will they do? What will they eat?"

It only takes a glance at the map of India to see that the country's northeast region occupies a unique place in the nation. Wedged up against Burma, Bhutan, Nepal, and China, it wraps around northern Bangladesh like an oversized arm around a child's shoulders. Joined with the rest of India through a "chicken neck" just thirteen miles wide, it was brought under Delhi's control by conquest from the east in 1826, when the British conquered Burma. Ethnically, the people are as much Tibetan or Burmese as they are South Asian, and many bristle at being part of India. The region's seven states are home to around a dozen

ethnic insurgencies, with causes ranging from greater auton-
omy within India to complete independence. Much of the
region remains off-limits to foreigners. Tourists are asked
to apply for special travel permits, which are only rarely
given.

Last winter, as Bangladesh was recovering from the
cyclone, I flew into the Indian northeastern city of Guwahati, on the banks of the Brahmaputra, and drove south
into the mountains. The city of Shillong, the regional capital under the British, lay just seventy-five miles away, but
the trip took us the entire morning. The road was pitted
and crowded. The driver of the car spent the whole time
on the horn, tapping lightly as he veered into a blind
curve, then blaring and braking to squeeze back between
oversized buses and wildly painted trucks.

In Shillong, we stopped to pick up Sanjeeb Kakoty, a
historian and documentary filmmaker, then continued
our drive towards India's border with Bangladesh. Kakoty
wore corduroy pants, brown leather sneakers, and a gray
collared shirt. He had a thin mustache, and his round face
was usually broken by a smile. We drove up a winding hill
road, past the town of Cherrapunji, said to be the wettest
place on Earth. The slopes on either side were steep. Fog
played in the ravines below. "Look at the hills here," said
Kakoty. "We're on a huge limestone rock. Years of deforestation have made it almost barren. No trees can grow.
Not even grass can grow. Just the rocks you see. On the
one hand, Cherrapunji has the highest rainfall in the
world. On the other, it doesn't have the trees or bushes to
break the rain. All the topsoil has been washed away. They
call it the wettest desert in the world."

The road ended in a parking lot and the entrance of a park. We paid our fee and walked inside. Teenagers picnicked on the grass. I could hear a radio playing American music from a stand of trees. Kakoty led the way across concrete paths to the edge of a cliff, ringed in metal fencing. The limestone outcropping stood out like the prow of a ship over the flat plains below. We were standing on the edge of India. Looking back at the curve of the plateau, I could see sheer drops and steep, forested slopes. In the haze below, snakes of glistening water flashed in the sun. "By definition, the plains means Bangladesh," said Kakoty. "Obviously, if that floods, they'll come up to the safe ground."

Until 1947, when the British abandoned their dreams of empire, the Bengali plains and the northeastern hills were part of the same colony. The border between India and what would become Bangladesh was drawn by a British bureaucrat named Cyril Radcliffe, who arrived in India for the first time just five weeks before the country was partitioned. He never visited the northeast, or indeed much of the rest of India. Asked to decide where the border south of Shillong would run, his commission drew the boundary at the bottom of the hills. Delhi's control began with the slopes. If the land was flat, it was given to East Pakistan (which would later become Bangladesh).

Migration from Bangladesh into India's northeast had been happening for centuries, but it had accelerated under the British, who coaxed Muslims from the Bengali plains into the predominantly Hindu region to work in the tea industry. During the Second World War, further resettlement was encouraged to put more land under cultivation and supply food to the front lines. The international border divided villages in two and separated farmers from their fields. It also changed the nature of migration. Until Radcliffe drew his line, a Bengali peasant could simply pick up and head north. Those that wanted to follow now faced crossing into another country.

"Tension is a constant piece of this area's political life," said Sanjib Baruah, a professor of political studies at Bard College in New York and a native of Assam. "And migration is never far away as a source of that tension." We spoke in a hotel restaurant in Guwahati over Indian beer and heavily seasoned peanuts. It's impossible to know just how many people have illegally moved from Bangladesh into India; new arrivals can easily blend into existing communities. Estimates range from the millions to the tens of millions. In the state of Assam, for instance, where immigration tensions are highest, Muslims make up the majority of six of the state's twenty-seven districts. "The demographic and social features of the entire western part of India's Assam state . . . have changed as a consequence of the influx of Bengali-speaking, predominantly Muslim refugees from Bangladesh," wrote Brahma Chellaney, a professor of Strategic Studies at the New Delhi–based Center for Policy Research. "It is perhaps the first time in modern history that a country has expanded its ethnic frontiers without expanding its political borders."

In 1979, tensions reached a crescendo as anti-immigrant feelings erupted into what became known as the Assam Agitation, six years of protests that all but shut down the state. The last big influx of immigrants had come eight years earlier when Bangladesh, backed by India, broke away from the rest of Pakistan, and the northeast was flooded by refugees. Most returned home once the fighting had stopped, but as many as one and a half million are thought to have stayed. The demonstrators, led by All Assam Students' Union, called for the deportation of illegal immigrants and boycotted elections they charged were determined by the votes of foreign nationals. "We're dealing with an area where there's an extremely poor documentation regime," said Baruah. "So, the dividing line between migration and citizenship is a thin one. Illegal immigration is almost de facto enfranchisement. It is changing the political balance, significantly. These are dangerous political waters."

Bridges were blown up. Police stations were stormed. Fearing more unrest, the government canceled the 1981 census in the state. The more radical protesters took up guns and headed for the hills, forming a militant movement that fights on today. "Illegal migration into Assam was the core issue behind the Assam student movement," wrote Srinivas Kumar Sinha, then the governor of Assam, in a report to the president of India. "It was also the prime contributory factor behind the outbreak of insurgency in the State." The message was clear. For India, immigration from Bangladesh had to be considered a matter of national security. "Protest against migration and the insurgency in Assam all started in the same year," said Baruah. "Managing this kind of migration would be very difficult for any society. Here it explodes into violence. In political instability. In insurgency."

As a political issue in the northeast, immigration remains without parallel. "Post-1979, almost all elections in Assam have been fought on the major agenda of who is for immigration, who is against immigration," said Kakoty. The agitation's outcome was an agreement between the demonstrators and the government that solved the problem on paper, but actually did very little. After all, the immigrants—whether old and legal or recently arrived—have the vote, and politicians aren't eager to offend them. Between 1986 and 2000, the police expelled fewer than 1,500 illegal immigrants. Demographics have continued to shift, much to the resentment of those who consider themselves native born. Every newspaper I picked up during my stay in the northeast referred to the issue, usually in the language of crisis. In conversation and in newspapers, the word used to refer to a Muslim of Bengali origin wasn't *immigrant* or *foreigner.* It was *infiltrator.*

Kakoty had turned his back to the Bengali plains and was leaning on the rail. "The government of Assam has been distributing land documents to all the immigrants," he said. "They have become bona fide settlers now. They own the land that they're sitting on. So what happens now? Probably the battle is lost. The war is not over. But they have come to stay."

"Six years of agitation produced nothing," he said. "The president of the student body became the chief minister. His deputy became the home minister. They couldn't do much to solve the problem. Initially, the insurgent movement had a lot of popular support. People called them their boys. That didn't work either. Now there's disappointment with the insurgent groups. People are asking, 'What's next?'"

"I remember the thinking in my own evolution," Kakoty said. "When I was a student, I would always think, 'Those people should be sent back.' I actually believed it could be done. But as you grow older, you realize it's not that easy. You can't chase away so many people." A hundred miles to our south, beyond a bulge of Bangladesh, lay the northeastern Indian state of Tripura, where immigrants from Bengal have long outnumbered the natives. The worry for the original people of Assam is that they too will become minorities in the lands their ancestors ruled. "I've become very liberal," said Kakoty. "Or maybe liberal is not quite the right word. I've become pragmatic. At one point in time I was very sure that we could chase them off. We could have a say over our own resources. Now I see, no, it's not going to happen."

"Very frankly, I've become a pessimist," he said. "I think our communities are doomed. We'll be absorbed by them, by Bangladesh and India."

On my way back from the border, I made a detour to Nellie, a roadside community of concrete buildings and palm trees along the national highway, where in February 1983 several nearby villages belonging to the Tiwa tribe organized themselves with machetes, bamboo spears, and poisoned arrows. Rumors had reached them of election violence in the capital, and they were determined to strike first. The killing was systematic. Two columns of tribesmen attacked early in the morning when the farmers were in the fields. They herded their victims along a canal, where a larger party waited in ambush. More than two thousand Bengali Muslims, mostly women and children, were killed.

Sanjoy Hazarika, reporting for the *New York Times*, arrived later that day, crossing the canal in a small dugout. "As I clambered up the small bank, my eyes were attacked by perhaps the most hideous scene that I have seen in all my years," he wrote in his book *Rites of Passage*. "An entire family was laid out at the top of the bank: parents and five children, of varying ages—the youngest was no older than an infant. Each one was dead, stabbed, slashed; the tiny one had been beheaded. Its head lay beside the body."

"I looked up and saw more bodies," he wrote. "I think that after that, we became numb to feeling—the paddy fields were full of young women, older women, old men, young children who were struck down. In one small patch of land, I counted two hundred bodies which lay where they had fallen. Yet, how had the young men survived?"

"The answer was simple," he wrote. "Because they could run faster than the women, the old, the infirm and the children."

I pulled up at the house of Kamal Patar, the village elder on whose lands the massacres took place. Patar is seventy-two years old, and I had come unannounced. As I waited for him to get ready, I sat in his receiving room and examined the decor. Pictures of cows, landscapes, and Ganesh, Hinduism's elephant-headed god, adorned the walls, which were painted baby blue. An incense holder the size of a large table lamp stood next to the armchair where I sat. Patar was barefoot when he arrived. He wore a white cotton shirt and a white cloth tied around his waist. Despite the heat, he had put on a woven, brown cotton vest. His face was deeply tanned, darker on his forehead and cheeks.

When he crossed his legs, I noticed the muscles in his calves. He ran a nonprofit group dedicated to development and reconciliation between his people and the Bengali-speaking Muslims. Every day, he would bicycle to the villages in which it operated. "More than fifteen hundred people died here," were the first words he said to me.

Throughout our conversation, which was held mostly through an interpreter with bits of broken English thrown in, Patar carefully drew a distinction between the Bengali-speaking Muslims in his village and the ones who had entered more recently. As we spoke, I referred to immigrants—whether old or new—as "Bangladeshis," a loaded term that implied their illegitimacy. Patar refused to follow suit, sticking instead to a more neutral term for his neighbors: "Muslims who live here."

"These people, these settlers," he said, "they are very old, unlike the ones that have recently come in. They are the owners of their land. They are as much citizens of this country as I am." Nonetheless, Patar left little doubt that he felt his community was under threat. Since the massacre, a large mosque had risen along the highway in the center of town. Villages to the west and north had become predominantly Muslim. Women dressed in black burqas, and Bengali was the language of the markets and the streets. When I asked Patar what he would think if traders or government offices began posting signs in the Bengali script, I could see I had made him angry. "Speaking their own language that they have learned in Bangladesh is not acceptable," he said. "They have come to Assam, and they should speak Assamese." It was a sentiment I came across often in the northeast. Patar had accepted the status quo, but he was not ready to make any further concessions. "The scenario has to remain what it is now," he said.

Global warming will pummel Bangladesh with rising seas in the south and increasingly unpredictable rivers from the north. Climate change could disrupt the monsoons, dropping agricultural productivity and facilitating the spread of waterborne diseases. The country's leading climate change expert, A. Atiq Rahman, the executive director of the Bangladesh Centre for Advanced Studies, likes to propose that people in the rich world should make up for their emissions by taking in environmental refugees, hosting one Bangladeshi family for every ten thousand tons of carbon emitted. He's only half joking. The Intergovernmental Panel on Climate Change estimates that the country's rice production could fall by nearly 10 percent. The wheat harvest could be cut by a third. As the quality of life plummets in their country, the pressure on Bangladeshis to migrate will intensify. Cyclones and floods could spark unprecedented movements of environmental refugees. The impact will be felt worldwide, but communities like Nellie will be the first to receive them.

I didn't mention climate change to Patar, but I did ask what he thought would happen if emigration from Bangladesh were to accelerate. Would his people accept new arrivals? He didn't think so. "There could be violence," he said. "A plate of rice is meant for one person. Today we have two people. Okay, we can survive. But ten or fifteen people cannot take from the same plate. There will be hunger somewhere."

On my return to Guwahati, I put the same question to

Samujjal Bhattacharya, the advisor for the All Assam Students' Union, the organization that had led the agitation in the 1980s. "It's a question of identity," he answered. "We're becoming minorities in our own motherland. Within ten years, an illegal Bangladeshi will be the chief minister of Assam. By 2020, the Bangladeshis will grab the northeast. We are sitting on the volcano. We have only two paths. Either we will have to surrender before illegal Bangladeshis. Or we have to unite, and we have to fight back. We are not advocating violence, but at the same time we must protect our identity."

"It's the duty of the government to evict all Bangladeshis and the jihadist groups," he said. "For any incident—which we do not want—the government of India will be responsible."

BEFORE KASHMIR BECAME FAMOUS FOR ITS HELLISH VIOLENCE, the mountain valley was called paradise on Earth. High above the sweltering plains of India, it made an ideal retreat from the heat, first by the Mughal empire, which turned the city of Srinagar into its summer capital, and then by the British. Denied permission to build onshore, colonial administrators constructed floating houseboats on the city's lakes and whiled away their summers under Himalayan slopes.

When I landed, the beauty was barely evident. I was behind schedule, held up by fog and flurries that had closed the airport for two days. By the time I disembarked, the snow had mostly melted, but the city was cold and gray. Leafless trees rose like pillars by the side of the road.

I had picked up a chest infection in Delhi, and the cold air raked at my lungs.

Kashmir has all the trappings of an occupied country. Though India has tried to decrease the visibility of the military in the capital, the signs were everywhere: sandbagged sentry posts on the street corners; concertina wire on the bridges; scarves wrapped around the faces of soldiers on patrol. My hotel lay not far from a military base. At dawn, I'd hear the call from the mosques, answered minutes later by the brass of a bugle.

On the streets, people stared straight ahead. They moved in small groups of three or four. Kashmir's national costume is a long, gray woolen gown that stretches from the shoulders to the knees, and most people wore one. Their cloaks robbed them of body language. Their faces, it seemed to me, were carefully expressionless. In the most crowded areas, Indian soldiers or Kashmiri police stood on every corner.

India and Pakistan have fought three wars over the region—in 1947, 1965, and 1999—but the most destructive conflict has been the insurgency that has pitted Kashmiris against the central government. India doesn't disclose how many troops it has deployed in the region and along the cease-fire line with Pakistan, but estimates run from two hundred thousand to six hundred thousand—far more than the total that the United States had in Iraq during the 2007 troop surge. Human rights violations committed by Indian security forces and the militants, who receive support and training from Pakistan, include forced disappearances, rape, torture, and summary executions. Tens of thousands of militants, Indian troops, and civilians have lost their lives in the fighting.

• • •

In Srinagar, I visited a woman named Safia Azar. My interpreter and I left our shoes on the porch and stepped inside her reception room. A line of armchairs were set against the wall, but we sat on the floor on a flower-print carpet and pulled blankets over our legs. Safia offered us clay pots filled with burning coals for warmth, then tucked her knees into her Kashmiri cloak as she kneeled across from us.

Safia was thirty-four years old, with a tender, round face under a yellow scarf. Fifteen years earlier, her husband, a timber trader named Humayun, had been arrested at an Indian checkpoint as he was driving to his sister's home. Safia and Humayun had been married for two years and were the parents of a six-month-old baby. Safia stood, walked to a bookshelf, and returned with a framed photograph. In the picture, her husband wore a loose brown shirt. His legs were crossed casually. He had thick hair and a thin mustache. His broad face was slightly out of focus, but the photographer had caught him as he looked up and captured the breaking of a surprised, loving smile.

The last time anybody in Safia's family saw her husband was the night after he was arrested. A convoy of soldiers brought him by to search the building. "Our house was like an army garrison," said Safia. Everybody in the house, including Safia and her son, was herded into a room. Only Humayun's mother was allowed out. Later, she would tell Safia that her husband looked as if he had been beaten. "The soldiers would say, 'Give us his weapon. Where is his weapon?'" said Safia. "My mother-in-law said, 'He has no weapon.'" Humayun asked to see his son, but the officer in charge wouldn't allow it. The last words he spoke before the soldiers took him away were to his mother: "Save me."

Safia lived in two rooms that had belonged to her father-in-law and shared the house with her husband's four brothers and their families. She had worked briefly in an elementary school as a teacher and was employed part-time in a stationery shop. An uncle was helping to pay for her fifteen-year-old son's studies. A month after her husband's arrest, the army announced that Humayun had escaped from custody in the city of Jammu. Later, a stranger told her he had seen him in prison. Since then, she had heard nothing. She had spent almost all of her adult life married to a missing man. "If he was dead, I might have thought of going to my mother's house," she said. "Maybe I would have thought of remarriage. But instead he's disappeared."

Kashmiri custom and her family's embrace of martyr-dom prevented her from marrying again, from finding a man to support her son, from moving on with her life. "My husband was made to disappear," said Safia. "The punishment wasn't given just to him, but also to me and my child. If he would have been killed, he would have been one among many. Disappearing is more punishing. It's more cruel."

The causes driving the conflict in Kashmir have little to do with global warming, but rising levels of greenhouse gases could easily worsen the conflict. The region and indeed the entire continent will come under heavy pressure as climate change wreaks havoc on the water cycle. The melting glaciers of the Himalayas feed seven great rivers, supplying water for a region of billions of people, one in five of whom already lack access to safe drinking water. The Intergovernmental Panel on Climate Change predicts that decreases in the availability of fresh water in Central, South, East, and Southeast Asia, particularly in large river basins, will adversely affect more than a billion people by 2050. "Glacier melt in the Himalayas is projected to increase flooding and rock avalanches from destabilized slopes and to affect water resources within the next two to three decades," reads the panel's 2007 report. "This will be followed by decreased river flows as the glaciers recede."

As precipitation patterns change or become less predictable, countries dependent on monsoons will be challenged with managing supplies even as flooding rivers and rising seas threaten to turn their fields into lakes. Since many Asian countries funnel 90 percent of their water into agriculture, disruption in availability will also mean drops in food production. The climate change panel estimates that crop yields could fall by 30 percent in Central and South Asia. "Glaciers are very important for the dry season supply of water," said Monirul Mirza, an environmental scientist at the University of Toronto. "The progression of rainfall starts from the Brahmaputra Basin, and as it progresses gradually to the west, the amount of moisture in the air gets less and less. Western India and Pakistan get very little rainfall." While glacial runoff makes up less than 10 percent of the volume of the vast Brahmaputra, the rivers that feed Kashmir, western India, and Pakistan depend on melting ice for roughly 80 percent of their flow.

In Ladakh, the Himalayan region uphill from Kashmir, rising temperatures have begun to eat away at the glaciers.

"Wherever you have glaciers, you have a stream," said Nawang Rigzin Jora, a state parliamentarian from the region. "Wherever you have a stream, you have habitation. Agriculture is spring fed, glacier fed. When the glaciers are receding, when there is less snowfall, obviously agriculture will become less sustainable. Things are going to be very difficult in the days to come."

Before leaving India, I flew to the city of Jammu and drove along the national highway into the mountains. The Indian state of Jammu and Kashmir can be roughly divided into three regions: Ladakh, where Tibetan culture is dominant; the majority Muslim and secessionist Kashmir valley; and Jammu, a predominantly Hindu area at the foot of the mountains and the start of the Punjabi plains. The region we were crossing is among the quietest in the state, but the military presence was still obvious. Patrols of soldiers strung themselves along the shoulder. Long convoys of dark green trucks trundled down the road. The landscape was mountainous. Rocky rivulets cut into the scrubby slopes. Snow crusted in the fields. When the clouds broke, they revealed snow- and pine-streaked peaks.

For sixty years, the conflict in Kashmir has been framed as one about identity. The region is central to India's claims of being a secular state, a keystone the diverse and factious nation feels it can't afford to lose. To Pakistan, the valley is a predominantly Muslim region fighting for the right to join a nation formed in the name of Islam. But lurking behind both countries' rationales is a more strategic concern. Eighty percent of Pakistan's agriculture depends on rivers that originate in Kashmir.

In recent years, recurring water shortages in the country have led to shortfalls of grain. In 2008, flour shortages and rising food prices became an issue in Pakistan's elections. The government deployed thousands of troops to guard its wheat stores. For Pakistan's leaders, who refer to Kashmir as their country's "lifeline," leaving the region in Indian hands means ceding control of their waters to a country with which they have fought four wars. "This water issue between India and Pakistan is the key of the problem," said Mohammad Yusuf Tarigami, a parliamentarian from Kashmir. "Much more than any other political or religious concern, water is the key."

In his memoirs, Major General Akbar Khan, the Pakistani leader of the raiders who seized western Kashmir in 1947, explicitly cited control of the region's waters as a reason for the attack. Pakistan's "agricultural economy was dependent particularly upon the rivers coming out of Kashmir," wrote Khan. "What then would be our position if [all of] Kashmir was in Indian hands?"

In 1990, nine years before he seized control of Pakistan in a military coup, Pervez Musharraf, then a brigadier studying at the Royal College of Defence Studies in London, presented a dissertation in which he analyzed his country's relationship with India. "The argument differed from the public stance taken by the Pakistani government in the last fifty years," wrote Sundeep Waslekar, president of the Strategic Foresight Group, a Mumbai-based think tank. "The public debate has always focused on issues of terrorism, human rights, and the legality of accession. It has never linked the conflict to the rivers of Jammu and

Kashmir. The brigadier [Musharraf] was suggesting that the rivers hold the key to the solution."

It's a theme that resonates with the militants who do the actual fighting. "Water is an elemental kind of existential threat," said Praveen Swami, an Indian journalist who has specialized in the militancy. "Among Islamists in Pakistan, the feeling is very strong that the water in Kashmir is about the survival of Islam." In an article for the Indian newsweekly *Frontline,* Swami compiled excerpts from the country's jihadi press. "Pakistan's vast agricultural lands are extremely dependent upon the large amount of river water which originates in Kashmir," Swami quoted the militant Islamist journal *Ghazwa* as saying. "If India succeeds in depriving Pakistan of these vital water resources, nothing can stop Pakistan's agricultural lands from turning into a desert."

My destination was the Baghliar Dam, a large hydropower project on the Chenab River that has been a point of contention between India and Pakistan. The dam, which had been scheduled to come online in the summer of 2008, was expected to more than double Kashmir's supply of electricity—at least in the summer months. "We have a resource here that could turn Kashmir into Singapore," Nisar Ali, dean of social sciences at the University of Kashmir, had told me. "We have plenty of water."

"But I have an engineering complaint," Ali said. "Kashmir is a place where we depend on snowfall. Snow remains on the mountains; then it melts. From May onwards, our rivers are in full bloom; water levels go up; it generates more power. And then from September until April, the water level goes substantially down."

In 1960, in an attempt to address the water issues, India and Pakistan signed a treaty dividing up the six tributaries that form the Indus River. The three eastern branches that flow through Indian Punjab were given to India. Use of the water from the other three—flowing through Jammu and Kashmir—were reserved for Pakistan. A cap was set on how much land Kashmir could irrigate. Reservoir lakes were prohibited. Strict limits were set on how and where water could be stored. Hydropower plants would have to make do with the strength of the natural current, regardless of seasonal variations. "If you need one hundred megawatts, you have to build for three hundred," Ali said. "For eight months, the generation will be reduced by at least one-third." In the summer, Jammu and Kashmir's electrical supply is roughly three hundred megawatts. In the winter, it drops to around seventy megawatts. The average house spends roughly sixteen hours without power. The state is forced to spend millions of dollars buying power from the national grid. "Our government has estimated that we have twenty thousand megawatts in potential hydropower," said Ali. "Of that, we are just harnessing less than half a percent today."

Warming weather has changed the way water flows in the valley. Everybody I spoke with in Jammu and Kashmir described winters in past decades as being much colder. "In January, it used to be very rare to have rain in Kashmir," said Arjimand Hussain Talib, the author of a report on climate change for ActionAid, an international aid group. "But now it is common. At this time, you would have had a good amount of snow."

Lacking good meteorological data, Talib pieced together records from an assortment of government sources. Reports of road closures and accounts of compensation for snow-related damage backed up anecdotal evidence of much whiter winters. Areas where the drifts once averaged two or three feet are now bare. The region receives less than half the snowfall it got forty years earlier. When the snow does come, it rarely sticks. Glaciers are retreating or disappearing. The snow line has moved up. Snowpack that once lasted until summer runs off before the flowers have time to bud. Talib had just returned from a visit to Gulmarg, a ski resort about an hour from Srinagar. He had taken a ski lift up to the top, where he found the snow was melting. "From the houses, from the trees, it was all dripping," he said.

Talib found no evidence that the total amount of precipitation in Kashmir has changed, but people across the region reported that the levels of streams and rivers had dropped. Snow and ice are no longer acting as natural reservoirs, thawing just in time for the planting season. Nearly 70 percent of the springs and natural ponds that Talib surveyed were dry during the summer months. "Right now, they might have water," he said. "But our main growing season goes from May until November. By then, they are gone." In February 2007, melting snow combined with heavy rainfall to undermine the mountain slopes. Landslides buried the national highway—Srinagar's only connection with the rest of India—for twelve days. The runoff overfilled the Jhelum and Chenab rivers. Flash floods washed away four people. Water once used for summer irrigation and power production was flushing away in winter floods.

"We cannot develop industry, because there is no power," said Ali. "We cannot develop commercial agriculture. There's no power, so mechanization is not possible. Employment opportunities decline. Investment opportunities decline. It is this thing that led to a conflict. In the 1986 elections, you had allegations of rigging that triggered the armed struggle, but the seeds were already sown in economics. The trigger was political, but if the economics had been in line, it might have found nothing to explode."

The Indian officials in charge of the Baghliar Dam refused to let me visit the site, so I was left to view the project from the highway. It looked like a giant tooth, striated by rain, embedded in a narrow valley between steep canyon walls. An Indian military base sprawled through a nearby town included a spread of administration buildings, a hospital, and a helicopter pad. Pakistan opposes the hydropower scheme and other projects like it, fearing they will give India the capacity to sow instability by cutting back its water supplies or, in a military confrontation, unleash floodwaters into the country's fields. "In a warlike situation, India could use the project like a bomb," was how one journalist who covered the dam's construction put it.

For Kashmir, climate change will mean a drop in its already meager power production. In Pakistan, which uses 96 percent of its water for agriculture, the disruptions in the water supply could easily lead to famine. In 2000, the amount of water available in the country per person per year was nearly three thousand cubic meters, according to the Asian Development Bank. By 2007, it was hovering at just over one thousand cubic meters, the point beyond which water shortages "threaten food production, slow down the economy, and damage ecosystems."

Since 2000, water shortages have begun to strain relations between the Pakistani provinces of Punjab and Sindh, according to Strategic Foresight Group. The two regions have clashed over how to manage water shortages. During seasons of droughts, Sindh's share of irrigation water has been cut by more than 25 percent a year. "There are so many issues of ethnic divisions in Pakistan," said Nils Gilman, an analyst at Global Business Network, a strategic consultancy based in San Francisco. "Try to imagine throwing water control issues into that mix. That could become a much more toxic issue between the people down in the valley and the people up in the hills. The question is do we think the Pakistani army is going to be able to manage those kinds of internal tensions? I wouldn't bet a lot of money on that."

Climate change will exacerbate water tensions all over the world. Peru, Ecuador, Bolivia, and other Latin American countries depend on glacial runoff for their drinking water. Australia has been suffering crippling droughts. In the American West, disappearing snowpack combined with drought could lead to widespread shortages, especially during the summer months when demand is highest. The Colorado River could lose as much as a third of its flow in the next fifty years. One study gives even odds that by 2017 Lake Mead could drop too low to provide power to the Hoover Dam.

In Asia, among the hardest to be hit will be China, where the amount of water available per person is already less than a quarter of the world average. Beijing has begun diverting water running from the Tibetan Plateau into its western regions. The three hundred miles of tunnels will be one of China's most technically challenging feats and will cost more than the $25 billion Three Gorges Dam. Reports that China would include the upwaters of the Brahmaputra, floated by a group of retired Chinese officials in a book called *Tibet's Waters Will Save China*, sent India into a panic. Chinese authorities quickly disavowed the plan—as too expensive, too difficult, and too controversial—but the construction of a hydropower dam on the Chinese side of the border has caused ripples of concern in Delhi.

Densely crowded, widely impoverished, geographically exposed, and geopolitically critical, South Asia may be the region where climate change has the chance to spark the greatest catastrophe. The world has been able to allow the conflict in Darfur to smolder in the desperate hope that the tragedy will burn itself out. But what happens when the same forces are set loose in a region we can't afford to ignore?

In the case of sudden cataclysmic flooding in Bangladesh, the international community will have to cope with a humanitarian emergency in which tens of millions of refugees suddenly flee towards India, Burma, and on to

China and Pakistan. "You have to ask yourself, where does that population go?" said Anthony Zinni, the former head of the United States Central Command, which oversees military operations in East Africa, the Middle East, and Central Asia. "What's going to be the scope of the humanitarian problems that have to be handled? You have an entire nation of millions of people that basically could be affected. What security pressures do they put on India?"

The United States military has increasingly deployed in response to natural disasters. It took the lead during the 2004 Christmas tsunami in the Indian Ocean, when more than fifteen thousand military personnel, twenty-five ships, and ninety-four aircraft delivered water, food, and supplies just days after the waves hit. When an earthquake struck Pakistan later that year, the United States sent in more than a thousand personnel to assist with the relief effort. American helicopters delivered more than ten million pounds of supplies and evacuated more than fifteen thousand people.

As climate-driven disasters begin to strike harder and more frequently, the United States will have to decide to what extent it will want to respond and start to figure out how to go about doing it. "The tsunami response has been held up as a kind of prototype for the role of the United States in certain kinds of extreme disaster situations," said Peter Ogden, a national security analyst at the Center for American Progress. "Our military is the one with the capacity. Even though a lot of other countries pitched in, it was ultimately United States hardware that was capable of actually projecting itself quickly enough to perform that operation."

"What's your ability to operate in an area that has very weak infrastructure—ports, airfields, things that you need?" Zinni said. "What are going to be the security issues? How are you going to gain control if the institutions that provide security are gone or wiped out? You could see a small microcosm of that even here in the United States during Katrina. The police force collapsed. We had tremendous problems in terms of law-and-order issues. It's going to be exponentially worse in the third world environment. If their institutions fail, our responsibilities may go beyond a humanitarian mission and into total reconstruction."

While the United States may not choose to respond to every climate-driven emergency, there's little chance it will be able to sit idly by if disaster unfolds in Bangladesh on an unprecedented scale. As a major producer of greenhouse gases and perhaps the only country logistically capable of deploying into a maelstrom of floodwater and misery, the world's only superpower would come under heavy pressure internationally and domestically to respond there and elsewhere. The question would then become: are the armed forces prepared to deploy in a way that doesn't compromise their safety or their other responsibilities?

"Right now, the United States would not be able, simply from a practical standpoint, to provide a huge number of troops if it were necessary to respond to a tsunami-like situation," said Ogden. "Are we going to take people who have just returned from Iraq for a fourth deployment and send them back overseas? I think it's difficult to see how politically that happens." The response to the 2004 tsunami unfolded in an environment that was entirely welcoming to American involvement. But there's no guarantee the next deployment won't find elements eager to take potshots at their would-be rescuers. "These are issues you'd want to start thinking about," said Ogden. "What would that require? Are we training for those sorts of scenarios, to put that many troops on the ground, but to keep them safe while at the same time being able to operate with the civilian population?"

"We have to ask ourselves about our national guard, the relationship between domestic and international disaster response," said Ogden. "It's not as though the United States is going to be insulated from the impacts of climate change. We may have the capacity or the resilience by virtue of our wealth and technology to respond, but that's only the case so long as we have that on hand. We can't afford to not be ready at home."

Nor will the threats from climate change be confined to the humanitarian. The standoff over Kashmir became especially perilous when Pakistan and India tested nuclear weapons in 1998. General Anthony Zinni was in charge of the United States Central Command the following year when an advance by Pakistan into Kashmir's glaciers nearly broke into all-out war. "It was beginning to escalate, almost World War One like," said Zinni. "They were on automatic on their mobilization back-and-forth." The stakes were high. The two sides hold limited numbers of weapons. A nuclear war would not guarantee mutually assured destruction. Response times would be measured in minutes. In any confrontation, there would be a high temptation to strike first. "You also have a conventional force mismatch," said Zinni. "You might drive Pakistan to a quick decision in a use-it-or-lose-it situation. You're not going to have the luxury of time if things explode." In 2000, Bill Clinton, then president of the United States, described the cease-fire line that divides Kashmir as "the most dangerous place in the world."

Weakened by tensions and humanitarian disasters from flooding and famine in Bangladesh, crowded upon by China, with its own water supplies under threat, India will face an increasingly desperate and dangerous neighbor. The Indus Water Treaty has survived three wars and nearly fifty years. It's often cited as an example of how resource scarcity can lead to cooperation rather than conflict. But its success has depended on the maintenance of a status quo that will be disrupted as the world warms: As long as India didn't obstruct the rivers, Pakistan was able to irrigate its fields. "The Indus Basin is going to be one of the earliest affected by climate change," said B. G. Verghese, an expert on water resources at the Centre for Policy Research in Delhi. "Pakistan will need to have more reservoirs. The storage sites in Pakistan are very limited, and they've almost exploited what they have. Further sites lie in territory controlled by India."

As the glaciers melt and the rivers dry, Pakistan—unstable, facing dramatic water shortages, caged in by India's vastly superior conventional forces—will be forced to make one of three choices. It can let its people starve. It can cooperate with India in building dams and reservoirs in Kashmir, handing over control of its waters to its enemy. Or it can somehow ramp up support for the insurgency, in a risky gamble that the militancy can bleed India's resolve without sparking a war that could quickly spiral out of control. "The idea of ceding territory to India is anathema," said Sumit Ganguly, a professor of political science at Indiana University who has studied the conflict. "Suffering, particularly for the elite, is unacceptable. So what's the other option? Escalate."

"It's very bad news," he said. "It's extremely grim."

EPILOGUE

On a rainy day in eastern Uganda, I followed a group of farmers up a steep mountain slope. My wet shoes slipped in the red mud as we climbed single file, pushing trembling umbrellas into a sagging sky. It was one of the most beautiful places I'd ever seen. Cows grazed on alpine hillsides under big-leafed banana trees. Clouds broke across rolling pastures and steep wooded drops. The air hung with the heavy smell of soil.

Our hike ended at the edge of Uganda's Mount Elgon National Park, the site of a tree-planting project designed to combat global warming by pulling greenhouse gases from the atmosphere. A nonprofit group was reforesting the park's perimeter, earning carbon credits that it then sold to airline passengers looking to make up for their emissions. The revenues were being used to plant more trees. In theory, everyone was winning. The air was getting cleaner, travelers were feeling less guilty, and the Ugandans were getting a bigger park.

But in place of the forest stretched rows of newly planted maize and budding green beans. The farmers with whom I had climbed had once lived just inside the reforested area.

Angry that their fields had been taken, they had fought their expulsion with lawsuits and machetes. When the courts granted an injunction against further evictions, they had taken it as permission to clear the land they considered theirs. Beyond the worn concrete post that marked the border of the park lay a stubble of tortured stumps—all that was left of the trees meant to absorb carbon dioxide.

The biggest challenge in the fight against climate change will be the countless cases where efforts to reduce greenhouse gases run up against local interests. The farmers of Mount Elgon benefit as much as anybody from the mitigation of climate change. As citizens of a country that shares a border with Sudan, planting in an area prone to drought, living their lives on a continent particularly vulnerable to climatic shifts, they will be especially exposed to the ravages of a warming world. But the gains from reforestation are distributed across the entire globe, blunting the blows of climate change for everyone by an infinitesimal amount. The problem was that for the farmers the benefit was too small to offset the loss of their fields.

Their calculation was not so different from those we make when we buy a bigger car, catch a plane to the Caribbean, or crank up the air-conditioning. Except that while our decisions concern matters of luxury, the farmers were making choices about matters of survival. The carbon we produce adds imperceptibly to a burden shouldered by everyone on Earth. But the sacrifices we will have to make to cut back on greenhouse gases may seem to be ours alone.

· · ·

The consequences of global warming described in this book may be alarming, but they're not meant to be alarmist. Some of them have already come about, and others may be unavoidable. Even if we were to stop releasing carbon immediately, the Earth's temperature will continue to rise for decades as the climate seeks a new equilibrium. Global warming is putting the whole planet under pressure, and the first to have felt its effects have been those in places where even a small change was enough to make a big difference. Arctic ice is giving way to open ocean. Glaciers are melting faster. Stronger storms are spreading fear along the coasts. Ecosystems are crawling north and uphill. Areas once a bit too cold for tropical diseases are hosting epidemics.

Even in the earliest stages, the impacts of climate change will be overwhelmingly negative. We are by definition adapted to our environments, even the harshest ones. The Inuit rely on the frozen seas to reach their traditional hunting grounds. The founders of our coastal cities didn't plan for rising seas. The plains of Punjab became the breadbasket of South Asia precisely because their farmers could rely on a steady stream of glacial melt. The finest wines flow from vineyards planted for the perfect climate.

Global warming will hurt most in areas least able to adapt. The Netherlands and Bangladesh share a common challenge. Most of the land in both countries lies near or below sea level. But while the Dutch are buttressing their dykes and designing floating houses that will rise with a flood, all the Bangladeshis can do is prepare to flee for higher ground. Drought-stricken Australia is unlikely to collapse into conflict. Its government can afford to spend billions on wind- and solar-powered desalinization plants. Meanwhile, the Pakistanis, who rely on melting glaciers for their water supply, will be able to do nothing but suffer the shortages.

The drought that set the stage for the conflict in Darfur struck across a wide swath just south of the Sahara, but not all the afflicted countries are as vulnerable as Sudan to collapsing into conflict. Tensions between farmers and nomads are also rising in Ghana, but it's unlikely that country will see anywhere near the same level of violence. Failing rains alone didn't create the nightmare in Darfur. It also took a callous regime in Khartoum, a fractious and disorganized rebellion, and the inattention of the world. Similarly, melting polar ice didn't have to unleash a scramble for the north. Had the opening of the Arctic not coincided with the rising cost of commodities and the implementation of a treaty dividing up the oceans, the world might have chosen a more cooperative approach.

Africa, South America, and Asia will be hit harder than the United States, Australia, and Europe. In Darfur, a long drought was enough to fan a fire into an inferno. Other areas will face other triggers. Drops in river levels will lead to competition over water. Crops will fail as milder winters spare warm-weather pests. Rich countries will be better able to absorb shocks in the price of food, to respond when hurricanes batter their coasts, to adapt to water shortages, to contain outbreaks when diseases expand their reach. In the early stages, the developed world will feel the pressures of climate change primarily secondhand, as its battered victims sweep up on its shores.

· · ·

Pinning down exact dates and locations for these future events is like trying to predict which wave will knock over a sandcastle. Most of what we know about climate change comes from computer models of the Earth's atmosphere, land cover, and oceans. The best ones include everything known about the way the three interact. Under what conditions do clouds form? What percentage of the sun's rays get reflected by forest cover as compared with desert sands? What are the differences in heat conduction in open fields and city streets? How well do tree roots retain groundwater? On computer screens you can watch the months pass by. Precipitation and temperature patterns flail across continents like flags in a storm.

But there are limits to the models' precision. Even the most advanced are able to replicate the Earth only at resolutions of about sixty miles. Smaller weather patterns—clouds, rivers, eddies in the seas—need to be approximated, as do physical parameters like the reflectivity of sand or the speed with which the oceans conduct heat. The result is a suite of programs, each with its own peculiar flavor. When compared to historical data, one model might underestimate drying in the Amazon. Another might exaggerate temperatures in Central Asia or mischaracterize the rainfall over the Indian Ocean. Even if we were to begin with a model that replicated the Earth exactly, it would not, in the end, necessarily get things right. Much in the way tiny changes in how you hit a pinball decide whether it ends up in the gutter or racking up the jackpot, small differences in starting temperatures and rain patterns can yield vastly different results. It's the famous flap of a butterfly in Brazil. A few fractions of a degree colder in Tibet, and you might see drought in Australia.

To iron out these artifacts, scientists run multiple models several times and average the outcomes. These composite results provide a good idea of how the world will change. How will temperatures increase if we double the amount of carbon in the atmosphere? What will happen to the Asian monsoons? How high will the seas rise? But there's one source of uncertainty no amount of computer power can eliminate: How much carbon will we pump into the air? Will we bring our emissions under control?

In a sense, this book is an exercise in optimism. Just what global warming will mean for the world will depend on just how serious we become in fighting it. The Intergovernmental Panel on Climate Change, the Nobel Prize–winning coalition of scientists, predicts that depending on how much carbon we release, global temperatures will rise between two and eleven degrees Fahrenheit by the end of this century. In comparison, during the last ice age, when glaciers buried the Great Lakes and blanketed Europe with ice as far as northern France, the planet was roughly eleven degrees colder than it is now.

The economist Nicholas Stern has calculated that tackling climate change now could cost just 1 percent of global domestic production, compared with damages that could shrink the world's economy by up to a fifth. His ambitious proposal, widely seen as the best we're likely able to do, aims to curb warming at about three and a half degrees Fahrenheit, enough to bring about most of the changes laid out in this book. In other words, the scenarios described in the preceding chapters could turn out to be the best we can expect under even the most rosy of assumptions.

The level of warming Stern proposes is also considered to be the point beyond which global warming risks slipping out of our control. After that, the Earth starts to undergo changes that perpetuate the thermometer's climb. Warm soils decompose faster, releasing carbon dioxide and methane. Hotter oceans absorb less. Melting permafrost unleashes millennia's worth of methane trapped in frozen bogs. At the North Pole, white ice is already giving way to black water; the seas are absorbing more sunlight. With each shift, the feedbacks pile up, the warming accelerates, and there's little we can do to slow it down.

Exactly what's in store for us if the world passes the point of no return is beyond the scope of this book, but predictions range from distressing to terrifying. Of course, the planet is no stranger to wild swings in temperature. Some tens of millions of years ago, Greenland was home to alligators, giant insects, and tropical ferns. Even as recently as five hundred thousand years ago, the Arctic island may have been covered by lush forest. But it's worth reminding ourselves that humans developed extensive agriculture only about ten thousand years ago and that the millennia since then have been among the most stable the world has ever experienced. The average global temperature never made a sustained shift of much more than a degree up or down.

Some predict that if we don't act soon to reign in our emissions, sea levels could rise by as much as twenty yards by the end of this century, putting New York, London, San Francisco, and countless other coastal cities underwater. Greenland's melting glaciers could stall the warm waters of the Gulf Stream, plunging Europe under ice. The Sahara could jump the Mediterranean. Droughts could devastate the American West. Mass extinctions could sweep the globe as ecosystems fail to keep up with rapidly rising temperatures. "Unchecked climate change equals the world depicted by Mad Max, only hotter, with no beaches, and perhaps with even more chaos," reads a joint report by the Center for Strategic and International Studies and the Center for a New American Security. "While such a characterization may seem extreme, a careful and thorough examination of all the many potential consequences associated with global climate change is profoundly disquieting. The collapse and chaos associated with extreme-climate-change futures would destabilize virtually every aspect of modern life. The only comparable experience for many in the group was considering what the aftermath of a U.S.-Soviet nuclear exchange might have entailed during the height of the Cold War."

Some blame minor, natural shifts in the local climate for the fall of the Mayans, the Anasazi, and the Greenland Norse. While modern society is far more advanced and adaptable, we may be entering a period of flux the likes of which human civilization has never seen.

Unfortunately, the battle against climate change is unlikely to be won with measures that will pay for themselves. New technologies can shoulder some of the burden, but tackling the problem will also depend on changing behaviors. Market forces may have the power to spark innovation, but only if government regulation puts a true price on carbon. Costs will have to be high almost by definition, or else they won't make a difference. Outside the biggest metropolises, just about every building raised in the past fifty years has been built for a world in which we spend hours each week in our cars. Even if the price of gasoline soars, our cities simply aren't designed for mass transit.

Nor is reducing emissions simply a matter of driving less. Nearly everything we do, from turning on the microwave to buying a bigger house to cracking open a beer, is responsible for releasing carbon into the atmosphere. Roughly half of the electricity used in the United States is produced from burning coal. The fabrication of cement—required for buildings, roads, and sewers—produces substantial greenhouse gases. Mass production, the engine of globalization, relies on cheap fuel to transport goods over long distances. Even the food we eat depends on mechanized agriculture and oil-based fertilizers.

Our window of opportunity is small, not only because greenhouse gases are accumulating at dangerous levels, but also because the problem is increasingly going global. By 2015, the developing world is expected to produce more than half the world's emissions. Any solution will require cooperation from India, China, and other countries where holding emissions flat could mean acquiescing to poverty. For people in industrialized countries, cutting carbon means accepting limits to our lifestyles. For the world's poor, as I saw when I climbed Mount Elgon with the Ugandan farmers, tackling emissions can mean sacrificing their livelihoods, if not their lives.

Bringing the people of the developing world on board will entail an acknowledgment that they too have a right to drive cars, light their houses, and run their factories. It will also require that we act soon, and drastically. As proponents for the world's poor like to point out, industrialized countries are responsible for the bulk of the carbon in the atmosphere.

If the richest people on the planet won't make economic sacrifices to address the problem, what chance is there that the rest of the world will?

In the course of researching this book, I flew roughly sixty thousand miles, releasing about twelve tons of carbon dioxide into the atmosphere. In flights alone, I was responsible for emitting about nine times what the average person does in a year. Like the Amazon researchers who flew to a climate change conference in Bali, or the scientists running diesel engines on the Canadian icebreaker, I put my own needs first. The world may have opened its eyes to climate change, but we're far from taking effective action. We are like a doctor puffing a cigarette, flipping through a medical journal on lung cancer and hoping it won't happen to us.

And just as with smoking, the choices we make now are cementing themselves in our future. Carbon dioxide stays in the atmosphere for decades. The effects we're seeing now are largely the results of gases pumped into the air during the Kennedy administration. Global emissions haven't slowed. They're accelerating. As the British journalist Mark Lynas points out, each breath we take contains more carbon dioxide than any breath ever taken by any human in evolutionary history. Climate stresses are piling up, and they will make it harder to address the root causes of global warming. In the midst of drought, conflict, migratory tensions, international crises, and humanitarian disasters, what time will we have for the complicated challenge of cutting carbon? The greenhouse gases we release today shape the world of the future. We don't have the luxury of waiting for devastating disasters to scare us into action.

NOTES

The information in the preceding chapters was drawn primarily from interviews, but also includes material from books, research papers, and news accounts. The following list is not meant to be exhaustive, but I hope it will provide a starting point for readers looking for more.

INTRODUCTION

For fuller investigations into the debate over the existence of global warming or its impacts on the environment, readers might try Elizabeth Kolbert's *Field Notes from a Catastrophe: Man, Nature, and Climate Change* (Bloomsbury USA, 2006), Tim Flannery's *The Weather Makers: How Man Is Changing the Climate and What It Means for Life on Earth* (Atlantic Monthly Press, 2006), and Al Gore's *An Inconvenient Truth: The Planetary Emergency of Global Warming and What We Can Do About It* (Rodale, 2006).

Chris Mooney's *Storm World: Hurricanes, Politics, and the Battle Over Global Warming* (Harcourt, 2007) covers the debate. In *Climate Change: What It Means for Us, Our Children, and Our Grandchildren* (ed. Joseph F. C. DiMento and Pamela M. Doughman [MIT Press, 2007]), the chapter by Naomi Oreskes, "The Scientific Consensus on Climate Change: How Do We Know We're Not

Wrong," explains why nearly every scientist believes our emissions are warming the Earth.

1: "THINGS WILL BREAK LOOSE FROM THE HANDS OF THE WISE MEN"

This chapter has its origins in "The Real Roots of Darfur," an article I wrote for the *Atlantic* (April 2007). It also includes reporting I did for stories in *Time:* "Nightmare in the Sand," May 9, 2004, and Simon Robinson, "The Tragedy of Sudan," September 26, 2004.

Alex de Waal describes his encounter with Sheikh Hilal Abdalla in "Counter-Insurgency on the Cheap," *London Review of Books,* August 5, 2004, and in a book coauthored with Julie Flint: *Darfur: A Short History of a Long War* (Zed Books, 2008).

The history of the rising tensions in Darfur is described in "Darfur Rising: Sudan's New Crisis," March 25, 2004, a report by the International Crisis Group.

Tim Burroughs's description of Bahai appeared in Scott Baldauf, "Sudan: Climate Change Escalates Darfur Crisis," *Christian Science Monitor,* July 27, 2007.

The links between warming oceans and the drought in Darfur are drawn in Alessandra Giannini et al., "Oceanic Forcing of Sahel Rainfall on Interannual to Interdecadal Time Scales," *Science* 302 (2003), and in Michela Biasutti and Alessandra Giannini, "Robust Sahel Drying in Response to Late 20th Century Forcings," *Geophysical Research Letters* 33, L11706 (2006).

David D. Zhang et al. explore the correlation between average temperature and warfare in ancient China in "Climate Change and War Frequency in Eastern China over the Last Millennium," *Human Ecology* 35 (2007).

In "Impacts of Climate Change: A Systems Vulnerability Approach to Consider the Potential Impacts to 2050 of a Mid-Upper Greenhouse Gas Emissions Scenario" (January 2007), Nils Gilman, Doug Randal, and Peter Schwartz of Global Business Network lay out their approach for predicting the effects of climate change.

International Alert offers its list of potential hot spots in Dan Smith and Janani Vivekananda, *A Climate of Conflict: The Links Between Climate Change, Peace and War* (International Alert, 2007).

The comments by L. K. Christian, Ghana's representative to the United Nations, are recounted in a United Nations Security Council press release, "Security Council Holds First-Ever Debate on Impact of Climate Change on Peace, Security, Hearing Over 50 Speakers," April 17, 2007.

The links between environmental degradation and conflict are laid out by Thomas Homer-Dixon in *Environment, Scarcity, and Violence* (Princeton University Press, 2001) and *The Upside of Down: Catastrophe, Creativity, and the Renewal of Civilization* (Island Press, 2008).

In a 1998 report for the Canadian International Development Agency, Philip Howard looked at Haiti's troublesome environment: "Environmental Scarcities and Conflict in Haiti: Ecology and Grievances in Haiti's Troubled Past and Uncertain Future."

The fates of the villages now under Lac de Péligre are described by Tracy Kidder in *Mountains Beyond Mountains: The Quest of Dr. Paul Farmer, a Man Who Would Cure the World* (Random House, 2003).

The United Nations Environment Programme draws the connection between climate change and Darfur in its report "Sudan: Post-Conflict Environmental Assessment" (2007). United Nations Secretary General Ban Ki-moon lays out the case in an op-ed, "A Climate Culprit in Darfur," *Washington Post*, June 16, 2007.

2: "WE'RE THE FAR COUNTRY"

The perils for coral in a warming world are described in "Coral Reefs and Global Climate Change: Potential Contributions of Climate Change to Stresses on Coral Reef Ecosystems," a 2004 report by Robert W. Buddemeier, Joan A. Kleypas, and Richard B. Aronson for the Pew Center on Global Climate Change.

The Intergovernmental Panel on Climate Change's Working Group I summarizes the scientific consensus on hurricanes in chapter 3 of its report "Climate Change 2007: The Physical Science Basis" (2007). Sea level rise is covered in chapter 5.

The reinsurer Swiss Re analyzes the storms of 2004 in "Hurricane Season 2004: Unusual, but Not Unexpected" (2006), and the flooding of New Orleans is considered by the risk modeling firm RMS in "Hurricane Katrina: Profile of a Super Cat: Lessons and Implications for Catastrophe Risk Management" (2005).

The impact of global warming on policyholders is explored by Evan Mills et al. in "Availability and Affordability of Insurance Under Climate Change: A Growing Challenge for the U.S." (2005), and Mills looks at the fate of the industry in "Insurance in a Climate of Change," *Science* 309 (2005).

Warren Buffett presented his worries about global warming in a 1992 letter to the shareholders of his holding company, Berkshire Hathaway.

Michael Treviño, a spokesman for Allstate, defends his company's cancellations in Karen Breslau, "The Insurance Climate Change," *Newsweek*, November 13, 2007.

A citizen of Key West bemoans the loss of his community in Jennifer Babson, "Insurance-Rate Hike Causes a Windstorm of Anger and Action," *Miami Herald*, April 9, 2006.

Douglas Brinkley chronicles the hammering of New Orleans in *The Great Deluge: Hurricane Katrina, New Orleans, and the Mississippi Gulf Coast* (William Morrow, 2006).

The city's recovery is tracked by the Greater New Orleans Community Data Center in its comprehensive series of reports "The New Orleans Index."

Ivan Miestchovich told National Public Radio his predictions for the city's future in "New Orleans Suburbs Rise in Wake of Flood," broadcast on *Weekend Edition*, March 18, 2007.

3: "A SPECTACULAR BIT OF GROWTH AS TIMES GET HARD"

Abdi Salan Mohammed Hassan's harrowing journey from Mogadishu to Lampedusa is recounted by Jeff Israely in "The Desperate Journey," *Time,* December 14, 2003.

The United Kingdom's military strategists lay out their vision of the
future in the third edition of "Strategic Trends" (2006), a report
prepared by the Ministry of Defence's Development, Concepts
and Doctrine Centre. In the United States, eleven retired admirals and generals weigh in with "National Security and the
Threat of Climate Change" (2007), a report for CNA, a Virginia-
based think tank.

Christian Aid's report "Human Tide: The Real Migration Crisis"
(2007) projects an explosion of climate refugees by 2050.

Fabrizio Gatti recounts his experience in Lampedusa's detention
center in the Italian newsweekly *L'espresso:* "Io, Clandestino a
Lampedusa," October 8, 2005.

Ian Cobain's time as a member in Nick Griffin's political party is
described in "The *Guardian* Journalist Who Became Central London Organiser for the BNP," *Guardian,* December 21, 2006.

Crossovers between environmentalism and the far right are
described by Jonathan Olsen in *Nature and Nationalism: Right-
Wing Ecology and the Politics of Identity in Contemporary Germany*
(Palgrave Macmillan, 1999).

James Lovelock lays out his apocalyptic view of climate change in
The Revenge of Gaia (Perseus Books Group, 2007).

Rear Admiral Chris Parry made his comments before the Royal
United Service Institute. They were quoted in a London Sunday
Times article, "Beware: The New Goths Are Coming," June 11,
2006.

The specter of a Green Junta is raised by Peter Wells in "The Green
Junta: Or, Is Democracy Sustainable?" *International Journal of
Environment and Sustainable Development* 6, no. 2 (2007).

4: "AT A NEW FRONTIER"

The interaction between malaria and the settlement of the Amazon is outlined by Marcia Caldas de Castro et al. in "Malaria Risk on the Amazon Frontier," *Proceedings of the National Academy of Sciences* 103, no. 7 (2006), and by Burton H. Singer and Marcia Caldas de Castro in "Agricultural Colonization and Malaria on the Amazon Frontier," *Annals of the New York Academy of Sciences* 954 (2001).

The research from the Peruvian Amazon cited by Jonathan Patz was published in Amy Vittor et al., "The Effect of Deforestation on the Human-Biting Rate of *Anopheles Darlingi,* the Primary Vector of Falciparum Malaria in the Peruvian Amazon," *American Journal of Tropical Medicine and Hygiene* 74, no. 1 (2006).

The story of Merrill Bahe's collapse and the search for the virus that killed him is told by Denise Grady in "Death at the Corners," *Discover,* December 1993; and by Laurie Garrett in "The War Between Man and Microbe," *Independent,* September 10, 1995.

The correlations between plague and climate in Kazakhstan are outlined in Nils Christian Stenseth et al., "Plague Dynamics are Driven by Climate Variation," *Proceedings of the National Academy of Sciences* 103, no. 35 (2006).

The Intergovernmental Panel on Climate Change's Working Group II provides an overview of the impact of climate change on human health in chapter 8 of its report "Climate Change 2007: Impacts, Adaption and Vulnerability" (2007). The Center for Health and the Global Environment at the Harvard Medical School covers the subject in its report "Climate Change Futures: Health, Ecological and Economic Dimension," ed. Paul Epstein and Evan Mills (2005).

Rafaella Angelini's comments appeared in Elisabeth Rosenthal, "As Earth Warms Up, Tropical Virus Moves to Italy," *New York Times,* December 23, 2007.

Paul Epstein describes the modern emergence of diseases in "Climate

Change and Public Health: Emerging Infectious Diseases," *Encyclopedia of Energy*, vol. 1 (Elsevier, 2004).

William Ruddiman's theories on deforestation and the Little Ice Age are laid out in *Plows, Plagues, and Petroleum: How Humans Took Control of Climate* (Princeton University Press, 2005).

Daniel Nepstad presented his predictions for the future of the Amazon in "The Amazon's Vicious Cycles: Drought and Fire in the Greenhouse," a 2007 report for the World Wide Fund for Nature.

The links between malaria and poverty are explored by Jeffrey Sachs and Pia Malaney in "The Economic and Social Burden of Malaria," *Nature* 415, no. 7 (2002).

John Podesta and Peter Ogden's predictions for the impact of disease on poor countries appeared in "The Security Implications of Climate Change," *Washington Quarterly* no. 1 (2007).

5: "BEAUTIFUL COUNTRY"

Pascal Yiou's use of harvest records to track temperature is described in Isabelle Chuine et al., "Grape Ripening as a Past Climate Indicator," *Nature* 432 (2004).

Gregory Jones et al. summarize their research in "Climate Change and Global Wine Quality," *Climatic Change* 73 (2005), and in "Climate Change: Observations, Projections, and General Implications for Viticulture and Wine Production," a paper Jones originally presented at the Climate and Viticulture Congress in Zaragoza, Spain, April 10–14, 2007, and published in the proceedings of the congress.

Robert Parker was profiled by William Langewiesche in "The Million Dollar Nose," *Atlantic*, December 2000.

Kim Nicholas Cahill's research on climate and agriculture can be found in David B. Lobell et al., "Historical Effects of Temperature and Precipitation on California Crop Yields," *Climatic Change* 81 (2007), and in Lobell et al., "Impacts of Future Climate Change on California Perennial Crop Yields: Model Projections with Climate and Crop Uncertainties," *Agricultural and Forest Meteorology* 141 (2006).

A catastrophic scenario for the American and California wine industry appeared in M. A. White et al., "Extreme Heat Reduces and Shifts United States Premium Wine Production in the 21st Century," *Proceedings of the National Academy of Sciences* 103, no. 30 (2006).

6: "EVERYTHING IS LATE IN CHURCHILL"

Part of this chapter uses reporting I did for *Monocle* magazine for "Frozen Assets—Norway," a story published in April 2007.

Angus and Bernice MacIver paint a loving portrait of their hometown in *Churchill on Hudson Bay* (Churchill Ladies Club, 1982).

The Intergovernmental Panel on Climate Change's Working Group II covers the Arctic in chapter 15 of its report "Climate Change 2007: Impacts, Adaption and Vulnerability" (2007). A 2004 report by Arctic Climate Impact Assessment, "Impacts of a Warming Arctic," covers the region in depth.

Barry Lopez's paean to the polar north, *Arctic Dreams* (Vintage, 2001), explores the region's history.

Roger Swanson celebrated his successful crossing of the Northwest Passage in Douglas Belkin, "As Arctic Ice Melts, Northwest Passage Beckons Sailors," *Wall Street Journal,* September 13, 2007.

Canadian defense minister Bill Graham's explanation of why he went to Hans Island appeared in Alexander Panetta, "Hands Off Hans Island: Graham to Denmark," The Canadian Press, July 22, 2005.

Worries about international shipping and Arctic oil spills were spelled out by Mike Byers in "Canada Must Seek Deal with U.S.: Vanishing Ice Puts Canadian Sovereignty in the Far North at Serious Risk," *Toronto Star,* October 27, 2006.

7: "AN ELEMENTAL KIND OF EXISTENTIAL THREAT"

Sanjoy Hazarika relates his encounters with the Bangladeshi family and his reporting on the massacre in Nellie in *Rites of Passage: Border Crossings, Imagined Homelands, India's East and Bangladesh* (Penguin Books India, 2000).

The Intergovernmental Panel on Climate Change's Working Group II explores the impact of climate change on the Himalayan glaciers in chapter 10 of its report "Climate Change 2007: Impacts, Adaption and Vulnerability" (2007).

Sabihuddin Ahmed laid out his worries about the impact of flooding in Bangladesh in "For My People, Climate Change Is a Matter of Life and Death," *Independent*, September 15, 2006.

An analysis of the northeast's problems is provided by Sanjib Baruah in "Postfrontier Blues: Toward a New Policy Framework for Northeast India" (2007), a report for the East-West Center in Washington.

Brahma Chellaney outlines the potential impacts of climate change on South Asia in "Climate Change and Security in Southern Asia: Understanding the National Security Implications," *RUSI Journal* 152, no. 2 (2007).

The need to train American troops in disaster relief is predicted in Podesta and Ogden, "The Security Implications of Climate Change."

A good overview of the situation in Kashmir can be found in Navnita Chadha Behera, *Demystifying Kashmir* (Brookings Institution Press, 2007).

Major General Akbar Khan provides his version of the first war over the region in *Raiders in Kashmir: Story of the Kashmir War, 1947–48* (Pak Publishers, 1970). Musharraf's dissertation is described in a report by the Strategic Foresight Group, "The Final Settlement: Restructuring India-Pakistan Relations" (2005), which lists water as an important driver of the conflict. The quotes by the jihadi press were compiled by Praveen Swami in

"Plot against Peace," in the Indian newsweekly *Frontline.* The importance of water in the Kashmiri conflict is explained by Erin Blankenship in "Kashmiri Water: Good Enough for Peace?" a report prepared for the Pugwash Conferences.

Arjimand Hussain Talib explores the impact of climate change in Kashmir in a report for ActionAid: "On the Brink?: A Report on Climate Change and its Impact in Kashmir" (2007).

Pakistan's water challenges are described in the Asian Development Bank's "Asian Water Development Outlook 2007."

EPILOGUE

The first paragraphs of this chapter are based on reporting I did for "The Other Side of Carbon Trading," an article that ran in *Fortune* on August 30, 2007.

In *The Economics of Climate Change: The Stern Review* (Cambridge University Press, 2007), Nicholas Stern describes why it's cheaper to halt climate change than to undergo it.

William James Burroughs contrasts the relative quiet of the last eight thousand years with the dramatic climatic swings of previous eras in *Climate Change in Prehistory: The End of the Reign of Chaos* (Cambridge University Press, 2005).

The Center for Strategic and International Studies and the Center for a New American Security outline their vision of a Mad Max future in Kurt M. Campbell et al., "The Age of Consequences: The Foreign Policy and National Security Implications of Global Climate Change" (November 2007).

In *Six Degrees: Our Future on a Hotter Planet* (National Geographic, 2008), Mark Lynas spells out what successive rises in temperatures could mean for our planet and our civilization.

Peter Schwartz and Doug Randall explore the future under catastrophic, sudden climate change in their report for Global Business Network, "An Abrupt Climate Change Scenario and Its Implications for United States National Security" (2003).

ACKNOWLEDGMENTS

One doesn't write a book without incurring a long list of debts. In addition to those quoted in these pages who generously shared their time, expertise, and insights, a number of people deserve my gratitude. Don Peck, my editor at the *Atlantic,* shepherded a speculation about the origins of the conflict in Darfur into the article that gave birth to what you're reading. My agent, Elisabeth Weed at Weed Literary, provided the inspiration and the title for the book, and David Patterson, my editor at Henry Holt, started to shape it even before he had signed on.

I'd be remiss if I didn't thank Simon Robinson and Eric Pooley at *Time* for getting me out of Iraq and sending me to report on Darfur as that conflict was breaking out, and Andrew Tuck of *Monocle* for commissioning a story on Norway. Robert Friedman, for whom I reported for *Fortune* from Uganda, deserves special thanks as a teacher, mentor, editor, and friend.

Listing everyone who helped me during my travels would take several pages, but special thanks should be extended to Will Van Sant, Nancy Klingener, and Heather Carruthers in Florida, Bill Drew and Mike Spence in Manitoba, Bob Faris and Darci Powell in California, and Henry Jardine, Shantie Mariet D'Souza, Mark Sappenfield, Dileep Chandan, Wahid Bukhari, Altaf Hussain, Luv Puri, Dan Isaacs, and Laurie Goering in India.

Marc Lacey, Melinda Miles, and Anna Osborne provided valuable help in Haiti. Audrey and Denice Warren, Musa Eubanks, Veda Manuel, and Sarah Kanter Brown did the same in New Orleans. Bettina Menne, Keith Alger, Thomas Lovejoy, Luiz Herman Soarez Gil, Joana Gabriela Mendes dos Santos, Marcia Caldas de Castro, and Gary Chandler were of great assistance for my trip to Brazil. I'm thankful to Flavio Di Giacomo for paving my way in Lampedusa, and to Torbjørn Goa for doing the same in Norway. Riccardo Rosati, Lidija Markovic, Paul Kalu, Peter Cunliffe-Jones, Nicola Peckett, and Larry Lohman were of company and help in London. Franceso Zizola was a great companion on that first trip to the border with Darfur. Nils Gilman, Alessandra Giannini, and Kim Nicholas Cahill provided valuable comments on the text.

Finally, my family deserves particular mention both for the feedback they provided and for putting up with me during the research and writing of the book. Madeleine Lapointe, Ron David, and Sophie Faris provided moral support. Bill Faris laid a firm foundation of commas and confidence. My wife, Federica Bianchi, lived without her husband for a year, propped up my feverish body in Kashmir, and carved structure into the majority of chapters. Along with Wang Jingjing, she offered a home to come back to. Leonardo Geronimo Faris waited for a father who "works too much" to join him in playing with his trains. Thanks. I'll do my best to make it up to you.

INDEX

ABOUT THE AUTHOR

STEPHAN FARIS is a journalist who specializes in writing about the developing world. Since 2000, he has covered Africa, the Middle East, and China for publications including *Time, Fortune, The Atlantic,* and *Salon.* He has lived in Nigeria, Kenya, Turkey, and China. He now lives in Rome with his wife and four-year-old son.